STATICS STUDY PACK

Free-Body Diagram Workbook & Chapter Reviews

PETER SCHIAVONE

UNIVERSITY OF ALBERTA

Engineering Mechanics
STATICS

Bedford • Fowler

FIFTH EDITION

Upper Saddle River, NJ 07458

Vice President and Editorial Director, ECS: *Marcia J. Horton*
Acquisitions Editor: *Tacy Quinn*
Associate Editor: *Dee Bernhard*
Managing Editor: *Scott Disanno*
Production Editor: *Craig Little*
Supplement Cover Manager: *Kenny Beck*
Manufacturing Buyer: *Lisa McDowell*

Pearson Prentice Hall
Pearson Education, Inc.
Upper Saddle River, NJ 07458

Printed in the United States of America
10 9 8 7 6 5 4 3 2 1

ISBN 0-13-614002-5
978-0-13-614002-3

Pearson Education Ltd., *London*
Pearson Education Australia Pty. Ltd., *Sydney*
Pearson Education Singapore, Pte. Ltd
Pearson Education North Asia Ltd., *Hong Kong*
Pearson Education Canada, Inc., *Toronto*
Pearson Education de Mexico, S.A. de C.V.
Pearson Education—Japan, *Tokyo*
Pearson Education Malaysia, Pte. Ltd.
Pearson Education, Inc., *Upper Saddle River, New Jersey*

ENGINEERING MECHANICS: STATICS, Fifth Edition

Companion Website Access Card

Thank you for purchasing a new copy of *Engineering Mechanics: Statics,* Fifth Edition, by Bedford & Fowler with this accompanying Study Pack. The access code below provides access to the Companion Website where you will find resources including:

- Self-assessment materials including study questions and additional practice problems with solutions
- Math topic reviews
- Computational Mechanics examples and MATLAB and Mathcad tutorials

(*Note:* If your professor assigned PH GradeAssist for online homework, you will need to obtain a separate access code.)

To access the Bedford & Fowler, *Engineering Mechanics: Statics* website for the first time:

1. Go to *www.prenhall.com/bedford*
2. Click on the title *Engineering Mechanics: Statics & Dynamics*, Fifth Edition.
3. Click on the "Companion Website" link and select the "Register" button.
4. Use a coin to scratch off the gray coating below and reveal your student access code. ***Do not use a knife or other sharp object as it may damage the code.*

5. On the registration page, enter your student access code. Do not type the dashes. You can use lower or uppercase letters.
6. Follow the on-screen instructions. If you need help during the online registration process, simply click the **Need Help?** icon.
7. Once your personal Login Name and Password are confirmed, you can begin using the Companion Website.

To login to the Bedford & Fowler, *Engineering Mechanics: Statics* website for the first time after you register:

Follow steps 1 and 2 to return to the Companion Website link and follow the prompts for "Returning User" to enter your Login Name and Password.

Note to Instructors: For access to Instructor Resources, contact your Prentice Hall Representative.

IMPORTANT: The access code on this page can only be used once to establish a subscription to the Bedford & Fowler, *Engineering Mechanics: Statics & Dynamics* website. If this access code has already been scratched off, it may no longer be valid. If this is the case, you can purchase a subscription by going to the Companion Website link and selecting "Get Access."

Upper Saddle River, NJ 07458
www.prenhall.com

To get help with registration, contact the Prentice Hall Media Support Line: 1-800-677-6337 or visit *http://247.prenhall.com/mediaform.html*

Contents

Preface v

PART I

1 Basic Concepts in Statics 3

1.1 Equilibrium 3

2 Free-Body Diagrams: the Basics 5

2.1 Free-Body Diagram: Object Modeled as a Particle 5
2.2 Free-Body Diagram: Object Modeled as a Rigid Body 8

3 Problems 13

3.1 Free-Body Diagrams in Particle Equilibrium 15
3.2 Free-Body Diagrams in the Equilibrium of a Rigid Body 45

PART II

1 Section-by-Section, Chapter-by-Chapter Review of A. Bedford & W. Fowler's text: *Engineering Mechanics STATICS (5th Edition)* 77

2 Answers to Review Questions 148

Preface

This supplement is divided into two parts. Part I is a workbook which explains how to draw and use free-body diagrams when solving problems in Statics. Part II provides a section-by-section, chapter-by-chapter summary of the key concepts, principles and equations from A. Bedford & W. Fowler's text: *Engineering Mechanics STATICS (5th Edition)*.

Part I. Free Body Diagram Workbook

A thorough understanding of how to draw and use a free-body diagram is absolutely essential when solving problems in mechanics.

This workbook consists mainly of a collection of problems intended to give the student practice in drawing and using free-body diagrams when solving problems in *Statics*.

All the problems are presented as *tutorial* problems with the solution only partially complete. The student is then expected to complete the solution by "filling in the blanks" in the spaces provided. This gives the student the opportunity to *build free-body diagrams in stages* and extract the relevant information from them when formulating equilibrium equations. Earlier problems provide students with partially drawn free-body diagrams and lots of hints to complete the solution. Later problems are more advanced and are designed to challenge the student more. The complete solution to each problem can be found at the back of the page. The problems are chosen from two-dimensional theories of particle and rigid body mechanics. Once the ideas and concepts developed in these problems have been understood and practiced, the student will find that they can be extended in a relatively straightforward manner to accommodate the corresponding three-dimensional theories.

The workbook begins with a brief primer on free-body diagrams: where they fit into the general procedure of solving problems in mechanics and why they are so important. Next follows a few examples to illustrate ideas and then the workbook problems.

For best results, the student should read the primer and then, beginning with the simpler problems, try to complete and understand the solution to each of the subsequent problems. The student should avoid the temptation to immediately look at the completed solution over the page. This solution should be accessed only as a last resort (after the student has struggled to the point of giving up), or to check the student's own solution after the fact. The idea behind this is very simple:

We learn most when we ***do*** *the thing we are trying to learn.*

In other words, reading through someone else's solution is not the same as actually working through the problem. In the former, the student gains *information*, in the latter the student gains *knowledge*. For example, how many people learn to swim or drive a car by reading an instruction manual?

Consequently, since the workbook is based on ***doing***, the student who persistently solves the problems in the workbook will ultimately gain a thorough, usable knowledge of how to draw and use free-body diagrams.

Part II. Chapter-by-Chapter Summaries

This part of the supplement provides a section-by-section, chapter-by-chapter summary of the key concepts, principles and equations from A. Bedford & W. Fowler's text: *Engineering Mechanics STATICS (5th Edition)*. We follow the same section and chapter order as that used in the *text* and summarize important concepts from each section in easy-to-understand language. We end each chapter summary with a simple set of review questions designed to see if the student has understood the key concepts and chapter objectives.

This section of the supplement will be useful both as a quick reference guide for important concepts and equations when solving problems in, for example, homework assignments or laboratories and also as a handy review when preparing for any quiz, test or examination.

P. Schiavone

PART I

FREE-BODY DIAGRAM WORK BOOK

1

Basic Concepts in Statics

Statics is a branch of mechanics that deals with the study of objects in *equilibrium*. In everyday conversation, equilibrium means an *unchanging state* or a *state of balance*. Examples of objects in equilibrium include pieces of furniture sitting at rest in a room or a person standing stationary on the sidewalk. If a train travels at constant speed on a straight track, objects that are at rest relative to the train, such as a person standing in the aisle, are in equilibrium since they are not accelerating. If the train should start to increase or decrease its speed, however, the person standing in the aisle would no longer be in equilibrium and might lose his balance.

More precisely, we say that objects are in equilibrium only if they are at rest (if originally at rest) or move with constant velocity (if originally in motion). The velocity must be measured relative to a frame of reference in which Newton's laws are valid, which is called an **inertial reference frame**. In most engineering applications, the velocity can be measured relative to the earth.

In mechanics, real objects (e.g. planets, cars, planes, tables, crates, etc) are represented or *modeled* using certain idealizations which simplify application of the relevant theory. In this book we refer to only two such models:

- **Particle or Point in Space.** A *particle* has mass but no size/shape. When an object's size/shape can be neglected so that only its mass is relevant to the description of its motion, the object can be modeled as a particle. This is the same thing as saying that the motion of the object can be modeled as the motion of a *point in space* (the point itself representing the center of mass of the moving object). For example, the size of an aircraft is insignificant when compared to the size of the earth and therefore the aircraft can be modeled as a particle (or point in space) when studying its three-dimensional motion in space.
- **Rigid Body.** A *rigid body* represents the next level of modeling sophistication after the particle. That is, a rigid body is a collection of particles (which therefore has mass) which has a significant size/shape but this size/shape cannot change. In other words, when an object is modeled as a rigid body, we assume that any deformations (changes in shape) are relatively small and can be neglected. Although any object does deform as it moves, if its deformation is small, *you can approximate its motion by modeling it as a rigid body*. For example, the actual deformations occurring in most structures and machines are relatively small so that the rigid body assumption is suitable in these cases.

1.1 Equilibrium

1.1.1 Equilibrium of an Object Modeled as a Particle

An object is in equilibrium provided it is at rest if originally at rest or has a constant velocity if originally in motion. To maintain equilibrium of an object modeled as a particle, it is necessary and sufficient to satisfy Newton's first law of motion which requires the resultant force acting on the object (or, more precisely, the object's mass center) to be zero. In other words

$$\sum \mathbf{F} = \mathbf{0} \tag{1.1}$$

where $\sum \mathbf{F}$ is the vector sum of all the external forces acting on the object.

Successful application of the equilibrium equation (1.1) requires a complete specification of all the known and unknown external forces ($\sum \mathbf{F}$) that act on the object. The best way to account for these is to draw the object's *free-body diagram*.

1.1.2 Equilibrium of an Object Modeled as a Rigid Body

An object modeled as a particle is assumed to have no shape. Hence only external *forces* enter into the equilibrium equation (1.1).On the other hand, an object modeled as a rigid body is assumed to have mass *and* (unchanging) shape. Hence, both forces and moments need to be taken into account when writing down the corresponding equilibrium equations. In fact, an object modeled as a rigid body will be in equilibrium provided the sum of all the external forces acting on the object is equal to zero *and* the sum of the external moments taken about any point is equal to zero. In other words:

$$\sum \mathbf{F} = \mathbf{0} \tag{1.2}$$

$$\sum \mathbf{M}_O = \mathbf{0} \tag{1.3}$$

where $\sum \mathbf{F}$ is the vector sum of all the external forces acting on the rigid body and $\sum \mathbf{M}_O$ is the sum of the external moments about an arbitrary point O.

Successful application of the equations of equilibrium (1.2) and (1.3) requires a complete specification of all the known and unknown external forces ($\sum \mathbf{F}$) and moments ($\sum \mathbf{M}_O$) that act on the object. The best way to account for these is again to draw the object's *free-body diagram*.

2

Free-Body Diagrams: the Basics

2.1 Free-Body Diagram: Object Modeled as a Particle

The equilibrium equation (1.1) is used to determine unknown forces acting on an object (modeled as a particle) in equilibrium. The first step in doing this is to draw the *free-body diagram* of the object to identify the external forces acting on it. The object's free-body diagram is simply a sketch of the object *freed* from its surroundings showing *all* the (external) forces that *act* on it. The diagram focuses your attention on the object of interest and helps you identify *all* the external forces acting. For example:

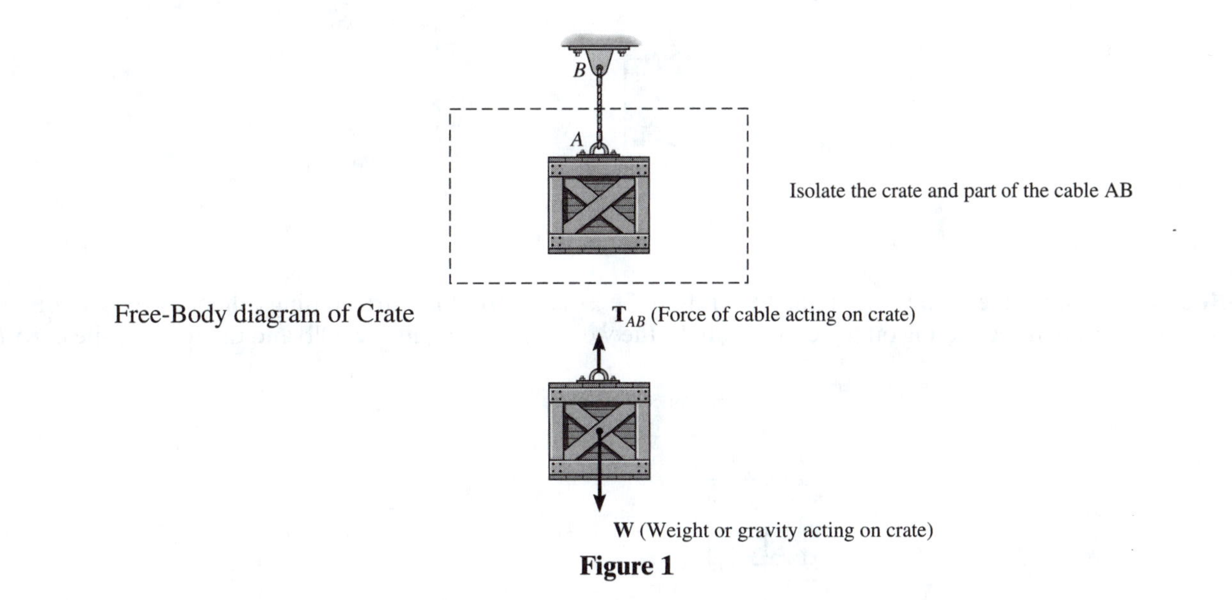

Figure 1

Note that once the crate is *separated* or *freed* from the system, forces which were previously internal to the system become external to the crate. For example, in Figure 1, such a force is the force of the cable AB *acting on the crate.*

Next, we present a formal procedure for drawing free-body diagrams for an object modeled as a particle.

2.1.1 Procedure for Drawing a Free-Body Diagram

1. *Identify the object you wish to isolate.* This choice is often dictated by the particular forces you wish to determine.
2. *Draw a sketch of the object isolated from its surroundings and show any relevant dimensions and angles.* Imagine the object to be isolated or cut free from the system of which it is a part. Your drawing should be reasonably accurate but it can omit irrelevant details.
3. *Show all external forces acting on the isolated object.* Indicate on this sketch *all* the external forces that act on the object. These forces can be *active forces*, which tend to set the object in motion, or they can be *reactive forces* which are the result of the constraints or supports that prevent motion. This stage is crucial: it may help to trace around the object's boundary, carefully noting each external force acting on it. Don't forget to include the weight of the object (unless it is being intentionally neglected).
4. *Identify and label each external force acting on the (isolated) object.* The forces that are known should be labeled with their known magnitudes and directions. Use letters to represent the magnitudes and arrows to represent the directions of forces that are unknown.
5. *The direction of a force having an unknown magnitude can be assumed.*

EXAMPLE 2.1

The crate in Figure 2 weighs 20lb. Draw free-body diagrams of the crate, the cord BD and the ring at B. Assume that the cords and the ring at B have negligible mass.

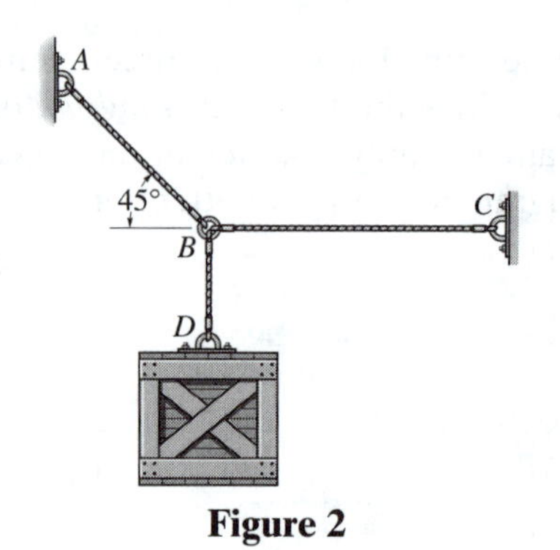

Figure 2

Solution

Free-Body Diagram for the Crate. Imagine the crate to be isolated from its surroundings, then, by inspection, there are only two external forces acting on the crate, namely, the weight with magnitude 20lb and the force of the cord BD.

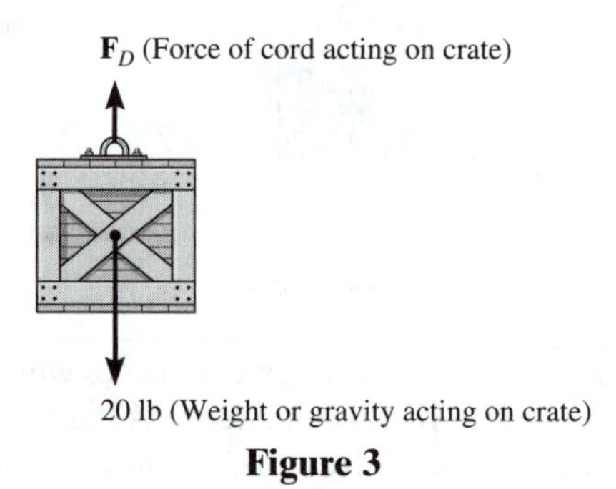

Figure 3

Free-Body Diagram for the Cord BD. Imagine the cord to be isolated from its surroundings, then, by inspection, there are only two external forces *acting on the cord*, namely, the force of the crate $\mathbf{F}_D$ and the force $\mathbf{F}_B$ caused by the ring. These forces both tend to pull on the cord so that the cord is in *tension*. Notice that $\mathbf{F}_D$ shown in this free-body diagram (Figure 4) is equal and opposite to that shown in Figure 3, a consequence of Newton's third law.

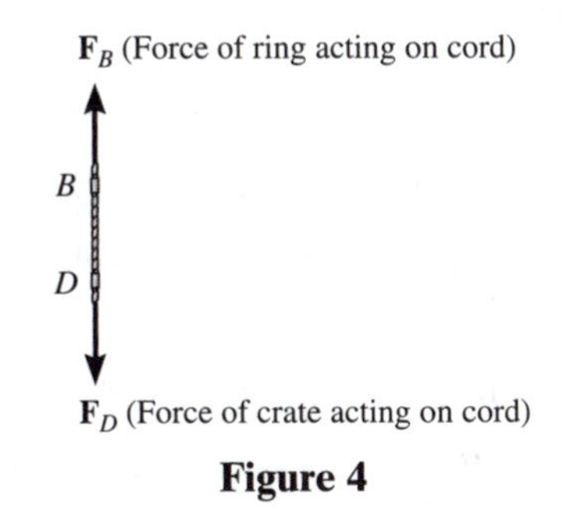

Figure 4

Free-Body Diagram for the ring at B Imagine the ring to be isolated from its surroundings, then, by inspection, there are actually three external forces acting on the ring, all caused by the attached cords. Notice that $\mathbf{F}_B$ shown in this free-body diagram (Figure 5) is equal and opposite to that shown in Figure 4, a consequence of Newton's third law. ◄

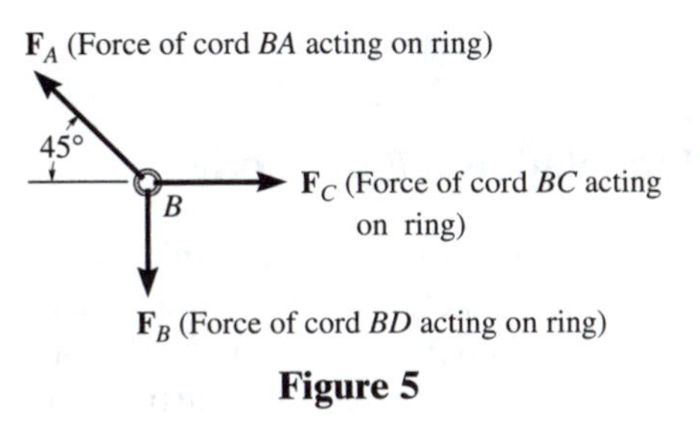

Figure 5

2.1.2 Using the Free-Body Diagram: Solving Equilibrium Problems

The free-body diagram is used to identify the unknown forces acting on the object when applying the equilibrium equation (1.1) to the object. The procedure for solving equilibrium problems is therefore as follows:

1. *Draw a free-body diagram*—you must choose an object to isolate that results in a free-body diagram including both known forces and forces you want to determine.
2. *Introduce a coordinate system* and establish the x, y-axes in any suitable orientation. Apply the equilibrium equation (1.1) in *component form* in each direction:

$$\sum F_x = 0 \text{ and } \sum F_y = 0 \tag{2.1}$$

   to obtain equations relating the known and unknown forces.
3. Components are positive if they are directed along a positive axis and negative if they are directed along a negative axis.
4. If more than two unknowns exist and the problem involves a spring, apply $F = ks$ to relate the magnitude F of the spring force to the deformation s of the spring (here, k is the spring constant).
5. If the solution yields a negative result, this indicates the sense of the force is the reverse of that shown/assumed on the free-body diagram.

EXAMPLE 2.2

In Example 2.1, the free-body diagrams established in Figures 3–5 give us a picture of all the information we need to apply the equilibrium equations (2.1) to find the various unknown forces. In fact, taking the positive x-direction to be horizontal ($\rightarrow +$) and the positive y-direction to be vertical ($\uparrow +$), the equilibrium equations (2.1) when applied to each of the objects (regarded as particles) are:

For the Crate: $\uparrow + \sum F_y = 0: \quad F_D - 20 = 0$ (See Figure 3)

$$F_D = 20 \text{ lb} \tag{2.2}$$

For the Cord BD: $\uparrow + \sum F_y = 0: \quad F_B - F_D = 0$ (See Figure 4)

$$F_B = F_D \tag{2.3}$$

For the Ring: $\uparrow + \sum F_y = 0: \quad F_A \sin 45^\circ - F_B = 0$ (See Figure 5) (2.4)

$\rightarrow + \sum F_x = 0: \quad F_C - F_A \cos 45^\circ = 0$ (See Figure 5) (2.5)

Equations (2.2) - (2.5) are now 4 equations which can be solved for the 4 unknowns F_A, F_B, F_C and F_D. That is: $F_B = 20$ lb; $F_D = 20$ lb, $F_A = 28.28$ lb, $F_C = 20$ lb. These are the magnitudes of each of the forces $\mathbf{F}_B$, $\mathbf{F}_D$, $\mathbf{F}_A$ and $\mathbf{F}_C$, respectively. The corresponding directions of each of these forces is shown in the free-body diagrams above (Figures 3–5). ◀

2.2 Free-Body Diagram: Object Modeled as a Rigid Body

The equilibrium equations (1.2) and (1.3) are used to determine unknown forces and moments acting on an object (modeled as a rigid body) in equilibrium. The first step in doing this is again to draw the *free-body diagram* of the object to identify *all of* the external forces and moments acting on it. The procedure for drawing a free-body diagram in this case is much the same as that for an object modeled as a particle with the main exception that now, because the object's "size/shape" is taken into account, it can support also external couple moments and moments of external forces.

2.2.1 Procedure for Drawing a Free-Body Diagram: Rigid Body

1. Imagine the body to be isolated or "cut free" from its constraints and connections and sketch its outlined shape.
2. Identify all the external forces and couple moments that act on the body. Those generally encountered are:
   (a) Applied loadings
   (b) Reactions occurring at the supports or at points of contact with other bodies (See Table 2.1)
   (c) The weight of the body (applied at the body's center of gravity G)
3. The forces and couple moments that are known should be labeled with their proper magnitudes and directions. Letters are used to represent the magnitudes and direction angles of forces and couple moments that are *unknown*. Establish an x, y-coordinate system so that these unknowns, for example, A_x, B_y etc can be identified. Indicate the dimensions of the body necessary for computing the moments of external forces. In particular, if a force or couple moment has a known line of action but unknown magnitude, the arrowhead which defines the sense of the vector can be assumed. The correctness of the assumed sense will become apparent after solving the equilibrium equations for the unknown magnitude. By definition, the magnitude of a vector is *always positive*, so that if the solution yields a *negative* scalar, the *minus sign* indicates that the vector's sense is *opposite* to that which was originally assumed.

Table 2.1. Supports used in Two-Dimensional Applications

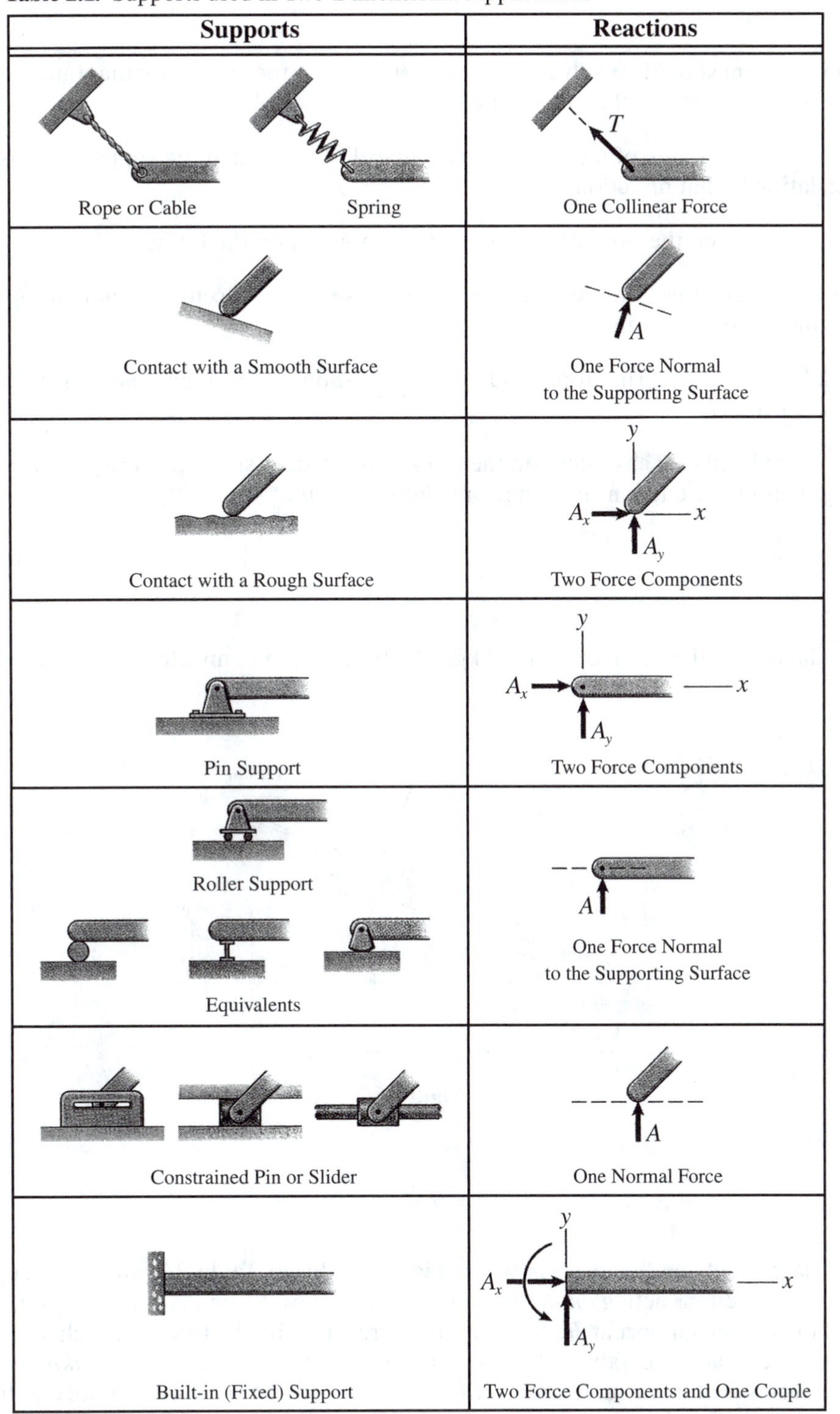

Supports	Reactions
Rope or Cable; Spring	One Collinear Force
Contact with a Smooth Surface	One Force Normal to the Supporting Surface
Contact with a Rough Surface	Two Force Components
Pin Support	Two Force Components
Roller Support; Equivalents	One Force Normal to the Supporting Surface
Constrained Pin or Slider	One Normal Force
Built-in (Fixed) Support	Two Force Components and One Couple

Important Points

- No equilibrium problem should be solved without first drawing the free-body diagram, so as to account for all the external forces and moments that act on the body.
- If a support *prevents translation* of a body in a particular direction, then the support exerts a force on the body to prevent translation in that direction.
- If *rotation is prevented* then the support exerts a couple moment on the body.
- Internal forces are never shown on the free-body diagram since they occur in equal but opposite collinear pairs and therefore cancel each other out.
- The weight of a body is an external force and its effect is shown as a single resultant force acting through the body's center of gravity G.
- *Couple moments* can be placed anywhere on the free-body diagram since they are *free vectors*. Forces can act at any point along their lines of action since they are *sliding vectors*.

Example 2.3

Draw the free-body diagram of the beam of mass 10 kg. The beam is pin-connected at A and rocker-supported at B.

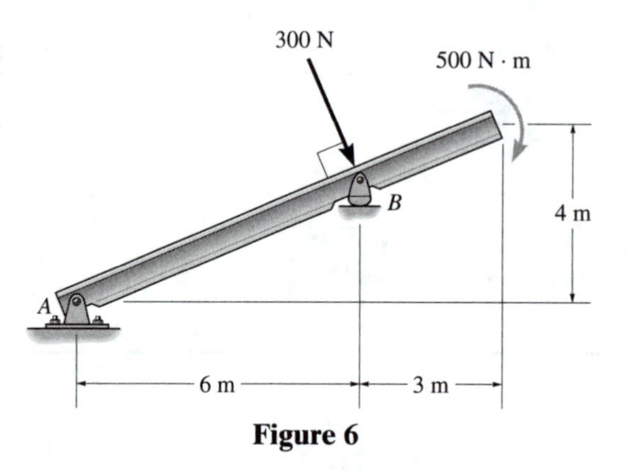

Figure 6

Solution

The free-body diagram of the beam is shown in Figure 7. From Table 2.1, since the support at A is a pin-connection, there are two reactions acting *on the beam at* A denoted by $\mathbf{A}_x$ and $\mathbf{A}_y$. In addition, there is one reaction *acting on the beam* at the rocker support at B. We denote this reaction by the force $\mathbf{F}$ which acts perpendicular to the surface at B, the point of contact (see Table 2.1). The magnitudes of these vectors are *unknown* and their sense has been *assumed* (the correctness of the assumed sense will become apparent after solving the equilibrium equations for the unknown magnitude i.e. if application of the equilibrium equations to the beam yields a negative result for the magnitude F, this indicates the sense of the force is the reverse of that shown/assumed on the free-body diagram). The weight of the beam acts through the beam's mass center. ◀

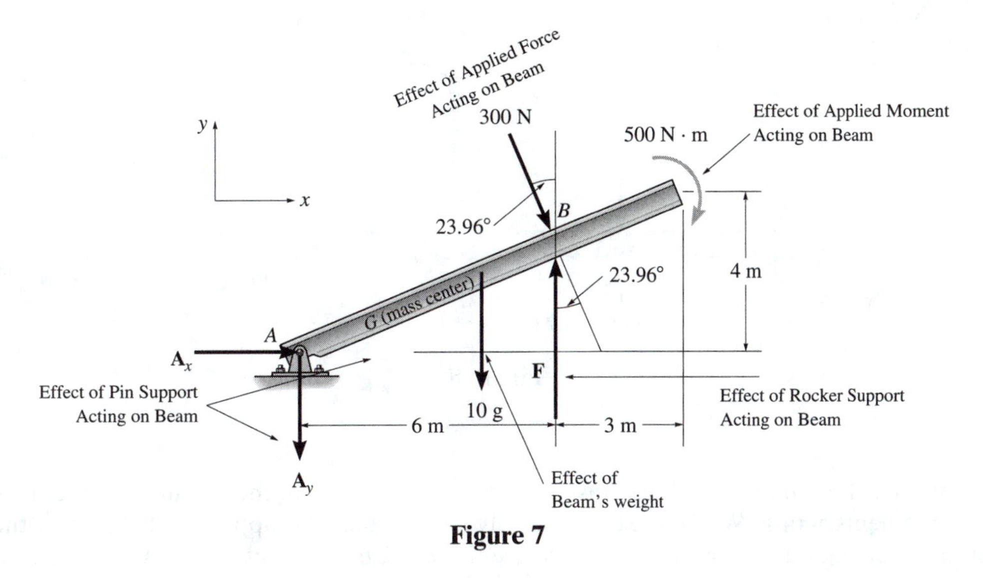

Figure 7

2.2.2 Using the Free-Body Diagram: Equilibrium

The equilibrium equations (1.2) and (1.3) can be written in component form as:

$$\sum F_x = 0, \tag{2.6}$$

$$\sum F_y = 0, \tag{2.7}$$

$$\sum M_O = 0, \tag{2.8}$$

Here, $\sum F_x$ and $\sum F_y$ represent, respectively, the algebraic sums of the x and y components of all the external forces acting on the body and $\sum M_O$ represents the algebraic sum of the couple moments and the moments of all the external force components about an axis perpendicular to the $x-y$ plane and passing through the arbitrary point O, which may lie either on or off the body. The procedure for solving equilibrium problems for a rigid body once the free-body diagram for the body is established, is as follows:

- Apply the moment equation of equilibrium (2.8), about a point (O) that lies at the intersection of the lines of action of two unknown forces. In this way, the moments of these unknowns are zero about O and a direct solution for the third unknown can be determined.
- When applying the force equilibrium equations (2.6) and (2.7), orient the x and y-axes along lines that will provide the simplest resolution of the forces into their x and y components.
- If the solution of the equilibrium equations yields a negative scalar for a force or couple moment magnitude, this indicates that the sense is opposite to that which was assumed on the free-body diagram.

Example 2.4

The pipe assembly has a built-in support and is subjected to two forces and a couple moment as shown. Find the reactions at A.

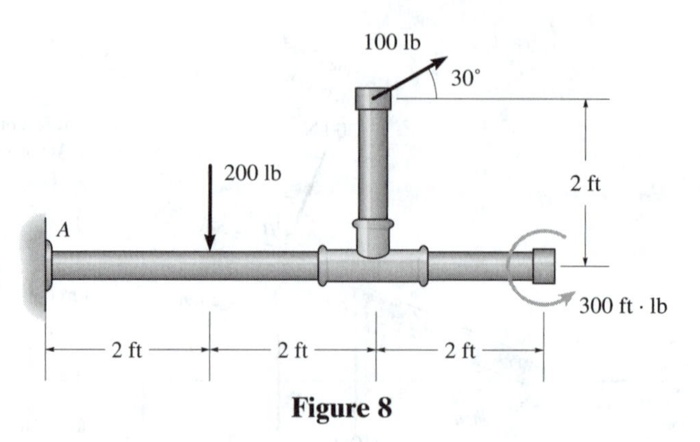

Figure 8

Solution

Free-Body Diagram The first thing to do is to draw the free-body diagram of the assembly in order to identify all the external forces and moments acting. We isolate the assembly from its built-in support at A (that way the reactions at A become external forces acting on the assembly). There are three unknown reactions at A: two force components A_x and A_y and a couple M_A (see Table 2.1). It might also be useful to resolve the applied 100 lb force into its components in anticipation of the application of the equilibrium equations (2.6)–(2.8). ◀

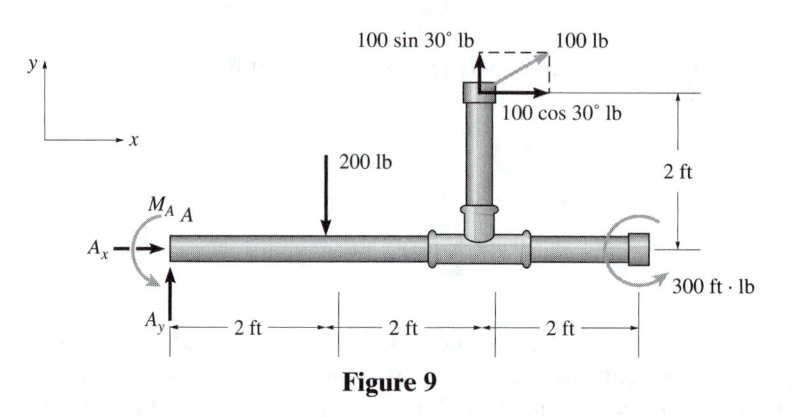

Figure 9

Equilibrium Equations The free-body diagram of the assembly suggests we can sum moments about the point A to eliminate the moment contribution of the reaction forces $\mathbf{A}_x$ and $\mathbf{A}_y$ acting on the beam. The equilibrium equations (2.6)–(2.8) are then:

$$\rightarrow +\sum F_x = 0: \quad A_x + 100\cos 30^\circ = 0$$
$$\uparrow +\sum F_y = 0: \quad A_y - 200 + 100\sin 30^\circ = 0$$

Taking counterclockwise as positive when computing moments, we have:

$$\sum M_A = 0: \quad M_A + 300 - (200)(2) - (100\cos 30^\circ)(2) + (100\sin 30^\circ)(4) = 0$$

(Notice that since the moment due to a couple is the same about any point, the moment about point A due to the 300 ft-lb counterclockwise couple is 300 ft-lb counterclockwise.) Solving these three equations we obtain the reaction components:

$$\boxed{A_x = -86.6 \text{ lb}, \qquad A_y = 150.0 \text{ lb}.}$$

(Note that we have obtained a negative value for A_x which means that the sense or direction of the force $\mathbf{A}_x$ is opposite to that which was assumed on the free-body diagram.

3

Problems

3.1 Free-Body Diagrams in Particle Equilibrium 15
3.2 Free-Body Diagrams in the Equilibrium of a Rigid Body 45

3.1 Free-Body Diagrams in Particle Equilibrium

In each of the following problems, assume that the objects are in equilibrium.

Problem 3.1.

The cables are used to support a block having a weight of 1000 lb. Draw a free-body diagram for the ring at A. The weight of the ring is negligible.

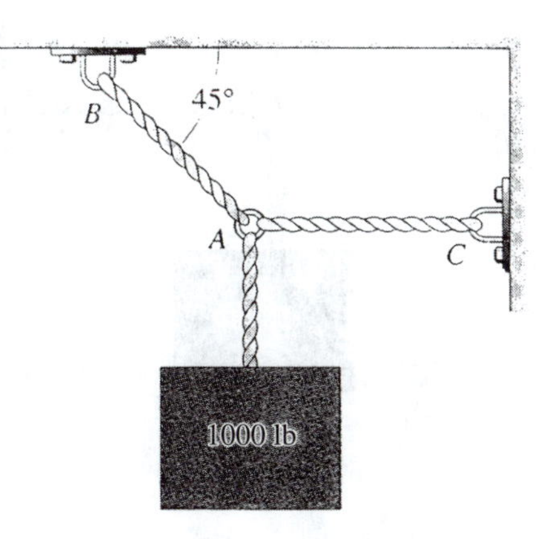

Solution

1. The ring acts only as a cable juncture and hence can be modeled as a particle (whose weight we are told to neglect).
2. Imagine the ring at A to be separated or detached from the system.
3. The (detached) ring at A is subjected to three *external* forces. They are caused by (remember to neglect the weight of the ring):

   i. **ii.**

   iii.
4. Draw the free-body diagram of the (detached) ring showing all these forces labeled with their magnitudes and directions. You should also include any other available information e.g. lengths, angles etc—which will help when formulating the equilibrium equations for the pulley.

A

Problem 3.1.

The cables are used to support a block having a weight of 1000 lb. Draw a free-body diagram for the ring at A. The weight of the ring is negligible.

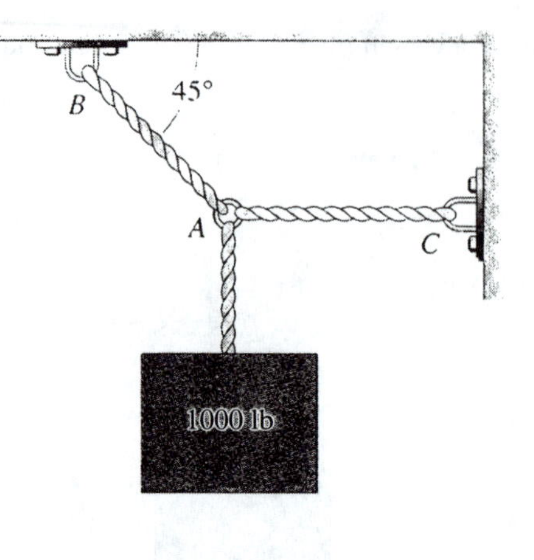

Solution

1. The ring acts only as a cable juncture and hence can be modeled as a particle (whose weight we are told to neglect).
2. Imagine the ring at A to be separated or detached from the system.
3. The (detached) ring at A is subjected to three *external* forces. They are caused by (remember to neglect the weight of the ring):
   - **i. CABLE** AB
   - **ii. CABLE** AC
   - **iii.** The 1000 lb **WEIGHT**
4. Draw the free-body diagram of the (detached) ring showing all these forces labeled with their magnitudes and directions. You should also include any other available information e.g. lengths, angles etc—which will help when formulating the equilibrium equations for the pulley.

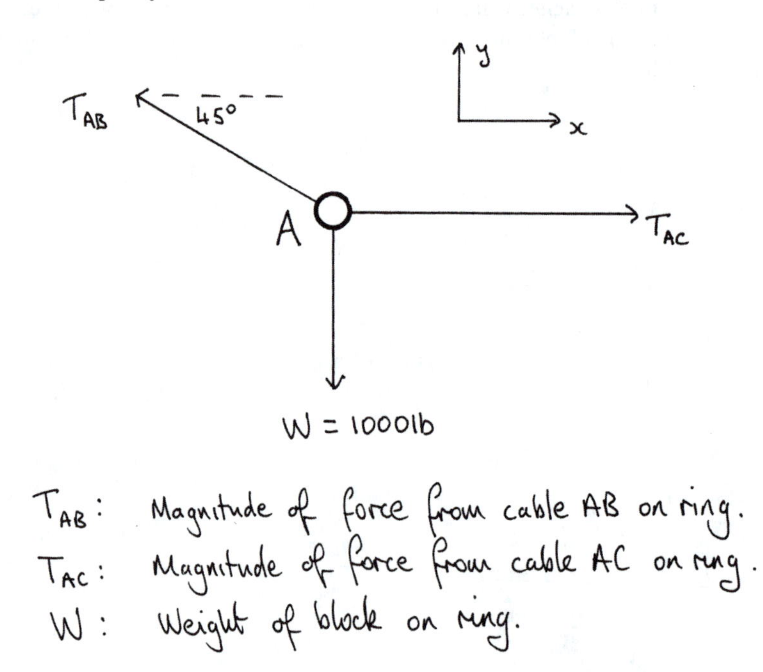

T_{AB}: Magnitude of force from cable AB on ring.
T_{AC}: Magnitude of force from cable AC on ring.
W: Weight of block on ring.

Problem 3.2.

The 200 lb horizontal bar is suspended by the springs A, B and C. The unstretched lengths of the springs are equal. The spring constants are k_A, k_B and k_C, respectively, with $k_A = k_C$. Draw a free-body diagram of the bar.

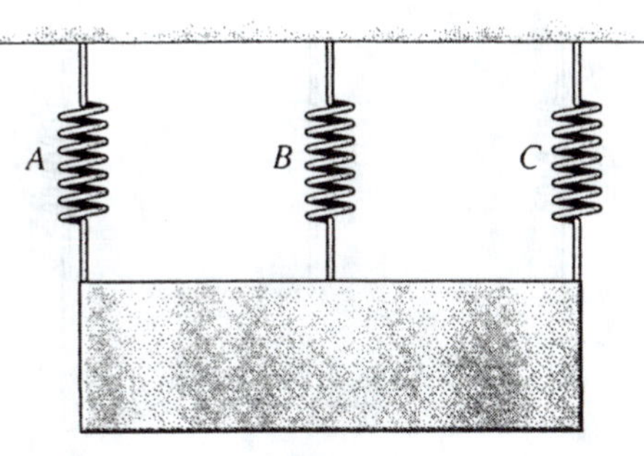

Solution

1. Imagine the bar to be separated or detached from the system.
2. The (detached) bar is subjected to four *external* forces. They are caused by:

   i. **ii.**

   iii. **iv.**

3. Draw the free-body diagram of the (detached) bar showing all these forces labeled with their magnitudes and directions. You should also include any other available information e.g. lengths, angles etc—which will help when formulating the equilibrium equations for the bar.

Problem 3.2.

The 200 lb horizontal bar is suspended by the springs A, B and C. The unstretched lengths of the springs are equal. The spring constants are k_A, k_B and k_C, respectively, with $k_A = k_C$. Draw a free-body diagram of the bar.

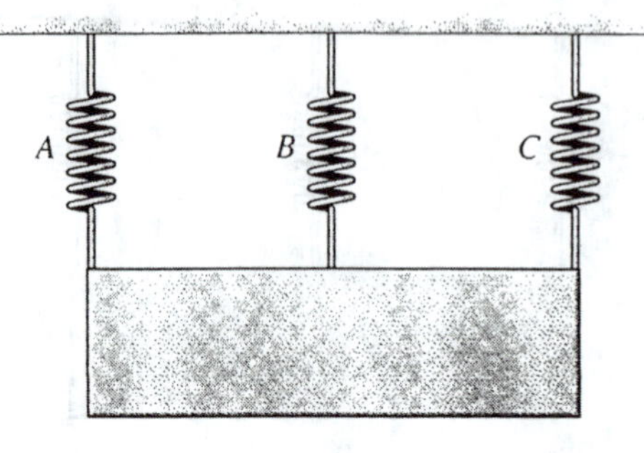

Solution

1. Imagine the bar to be separated or detached from the system.
2. The (detached) bar is subjected to four *external* forces. They are caused by:

   i. SPRING A

   ii. SPRING B

   iii. SPRING C

   iv. THE BAR'S WEIGHT (200 lb)
3. Draw the free-body diagram of the (detached) bar showing all these forces labeled with their magnitudes and directions. You should also include any other available information e.g. lengths, angles etc—which will help when formulating the equilibrium equations for the bar.

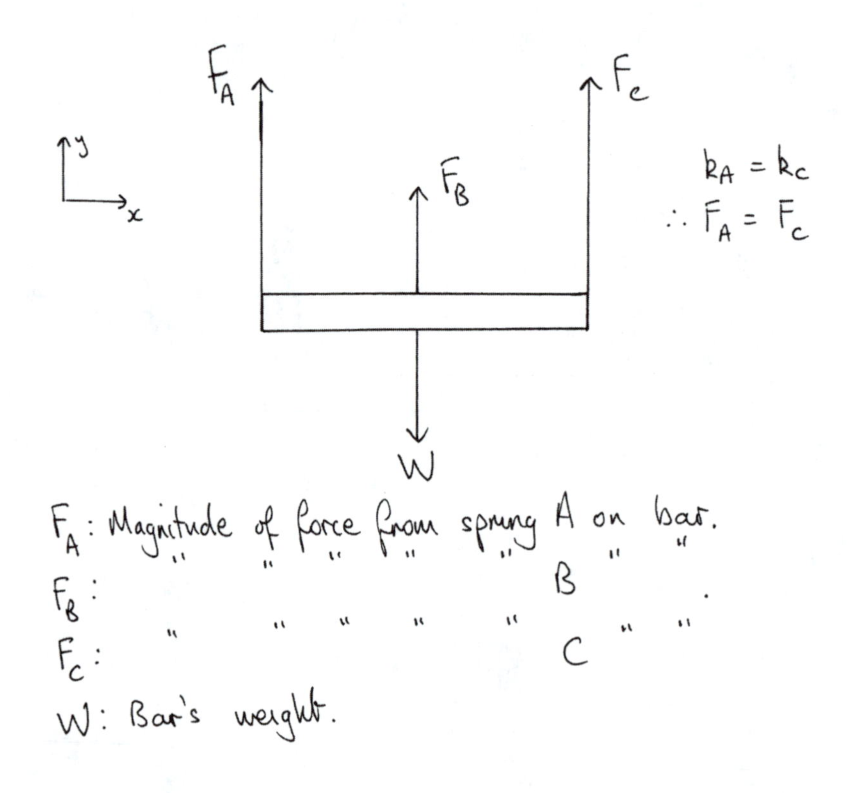

Problem 3.3.

The system of frictionless pulleys is in equilibrium. Draw separate free-body diagrams of the pulley at A and the bar at B. Neglect the weight of the pulleys.

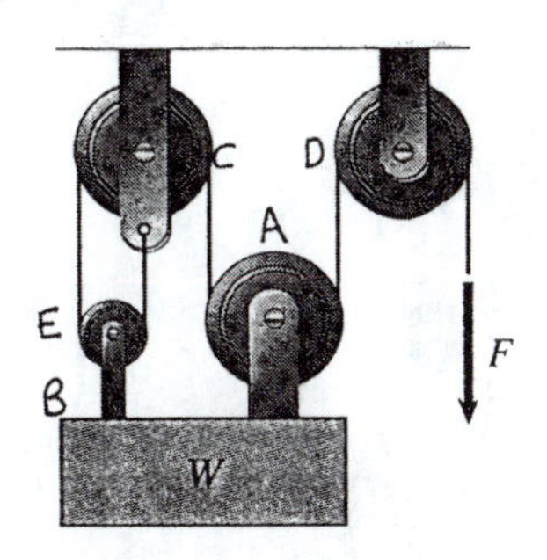

Solution

1. Imagine the pulley at A and the bar at B to be separated or detached from the system.
2. The (detached) pulley at A is subjected to three *external* forces.
   They are caused by (remember that a frictionless pulley changes the direction of a force but not its magnitude):

   i. **ii.**

   iii.

   The (detached) bar at B is subjected to three *external* forces. They are caused by:

   i. **ii.**

   iii.
3. Draw the free-body diagram of the (detached) pulley and the detached bar showing all these forces labeled with their magnitudes and directions. You should also include any other available information e.g. lengths, angles etc - which will help when formulating the equilibrium equations.

A B

Problem 3.3.

The system of frictionless pulleys is in equilibrium. Draw separate free-body diagrams of the pulley at A and the bar at B. Neglect the weight of the pulleys.

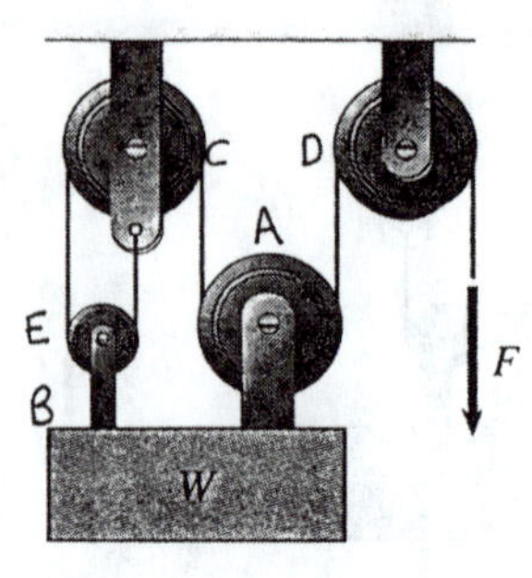

Solution

1. Imagine the pulley at A and the bar at B to be separated or detached from the system.
2. The (detached) pulley at A is subjected to three *external* forces. They are caused by (remember that a frictionless pulley changes the direction of a force but not its magnitude):
   - **i. CORD** AC
   - **ii. CORD** AD
   - **iii. REACTION OF WEIGHT ON PULLEY** A

   The (detached) bar at B is subjected to three *external* forces. They are caused by:
   - **i. REACTION OF PULLEY** A **ON WEIGHT**
   - **ii. REACTION OF PULLEY** E **ON WEIGHT**
   - **iii. THE WEIGHT** W
3. Draw the free-body diagram of the (detached) pulley and the detached bar showing all these forces labeled with their magnitudes and directions. You should also include any other available information e.g. lengths, angles etc - which will help when formulating the equilibrium equations.

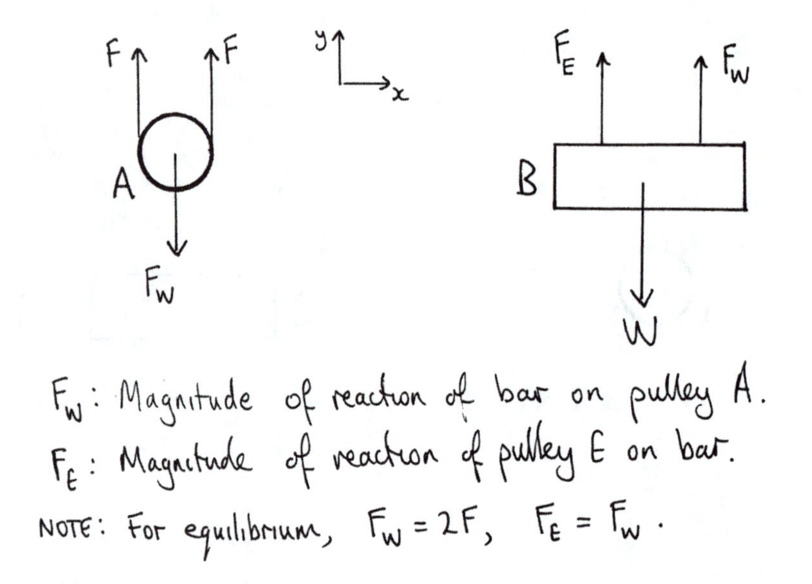

Problem 3.4.

The rocket is suspended by two cables. The mass of the rocket is 45 Mg. Draw a free-body diagram of the support collar at A.

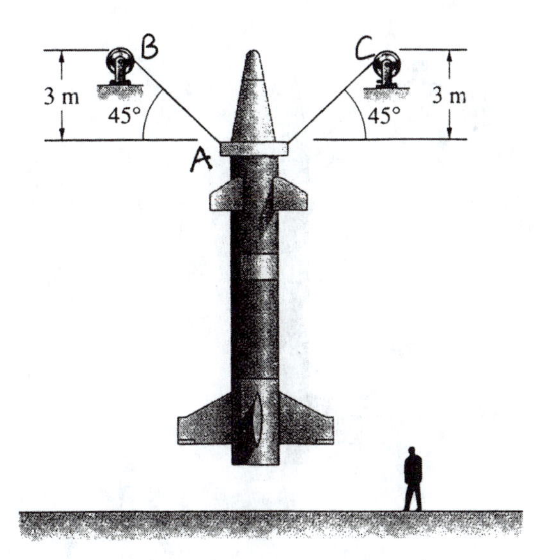

Solution

1. Imagine the collar at A to be separated or detached from the system.
2. The (detached) collar is subjected to three *external* forces. They are caused by:

   i. **ii.**

   iii.

3. Draw the free-body diagram of the (detached) collar showing all these forces labeled with their magnitudes and directions. You should also include any other available information e.g. lengths, angles etc—which will help when formulating the equilibrium equations for the pulley.

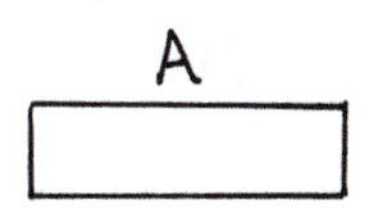

Problem 3.4.

The rocket is suspended by two cables. The mass of the rocket is 45Mg. Draw a free-body diagram of the support collar at A.

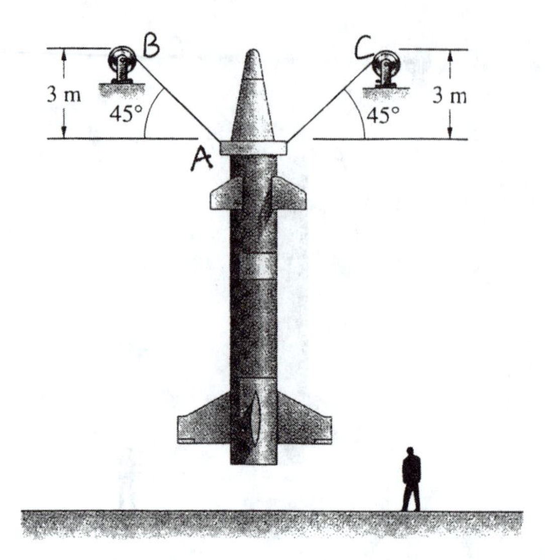

Solution

1. Imagine the collar at A to be separated or detached from the system.
2. The (detached) collar is subjected to three *external* forces. They are caused by:

   i. CABLE AB **ii. CABLE** AC

   iii. WEIGHT OF COLLAR

3. Draw the free-body diagram of the (detached) collar showing all these forces labeled with their magnitudes and directions. You should also include any other available information e.g. lengths, angles etc—which will help when formulating the equilibrium equations for the pulley.

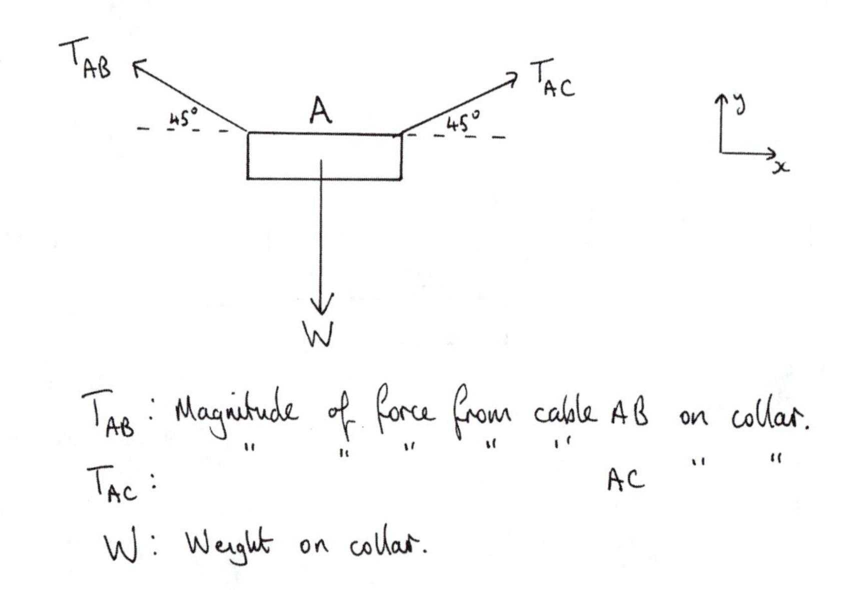

Problem 3.5.

A construction worker holds a 500-lb crate in the (equilibrium) position shown. Draw a free-body diagram for the cable juncture at A. Neglect the mass of the cable juncture.

Solution

1. Imagine the cable juncture to be separated or detached from the system.
2. The (detached) cable juncture is subjected to three *external* forces. They are caused by:

   i.　　　　**ii.**

   iii.

3. Draw the free-body diagram of the (detached) juncture showing all these forces labeled with their magnitudes and directions. You should also include any other available information e.g. lengths, angles etc—which will help when formulating the equilibrium equations.

A

Problem 3.5.

A construction worker holds a 500-lb crate in the (equilibrium) position shown. Draw a free-body diagram for the cable juncture at A. Neglect the mass of the cable juncture.

Solution

1. Imagine the cable juncture to be separated or detached from the system.
2. The (detached) cable juncture is subjected to three *external* forces. They are caused by:

   i. TENSION IN CABLE AB

   ii. FORCE EXERTED BY WORKER THROUGH CABLE AC

   iii. WEIGHT OF CRATE

3. Draw the free-body diagram of the (detached) juncture showing all these forces labeled with their magnitudes and directions. You should also include any other available information e.g. lengths, angles etc—which will help when formulating the equilibrium equations.

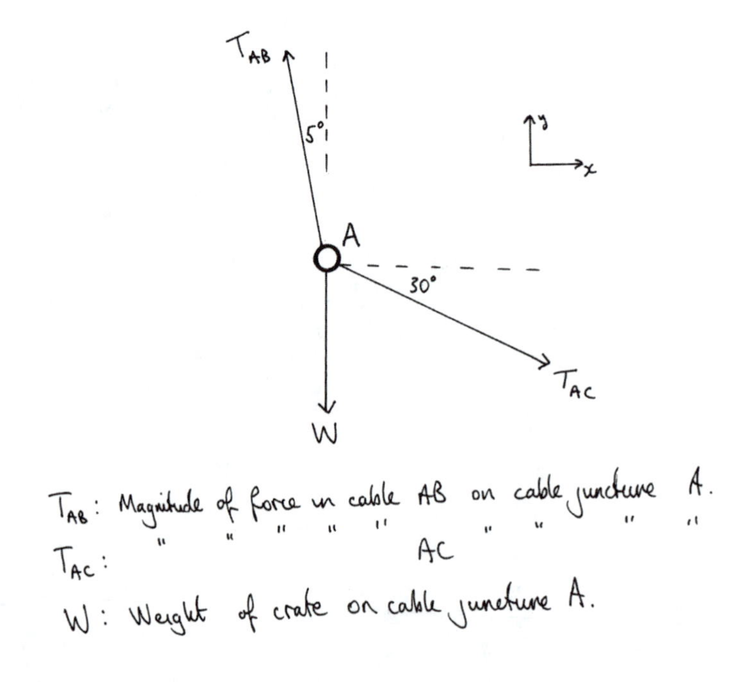

T_{AB}: Magnitude of force in cable AB on cable juncture A.
T_{AC}: " " " " " AC " " " "
W: Weight of crate on cable juncture A.

Problem 3.6.

The steel ball has a weight of 100 lb and is being hoisted at uniform velocity. The system is held in equilibrium at angle θ by the appropriate force in each cord. Draw the free-body diagram for the *small* pulley.

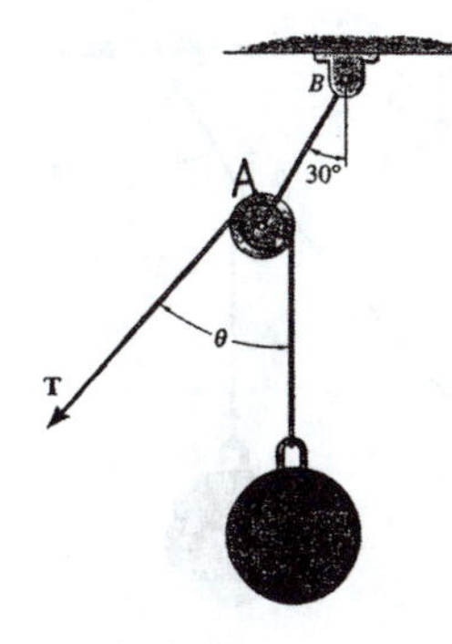

Solution

1. The pulley has *negligible shape* (we are told it is *small*) so that it can be modeled as a particle.
2. Imagine the pulley to be separated or detached from the system.
3. The (detached) pulley is subjected to three *external* forces. They are caused by:

   i. **ii.**

   iii.

4. Draw the free-body diagram of the (detached) pulley showing all these forces labeled with their magnitudes and directions. You should also include any other available information e.g. lengths, angles etc—which will help when formulating the equilibrium equations for the pulley.

Problem 3.6.

The steel ball has a weight of 100 lb and is being hoisted at uniform velocity. The system is held in equilibrium at angle θ by the appropriate force in each cord. Draw the free-body diagram for the *small* pulley.

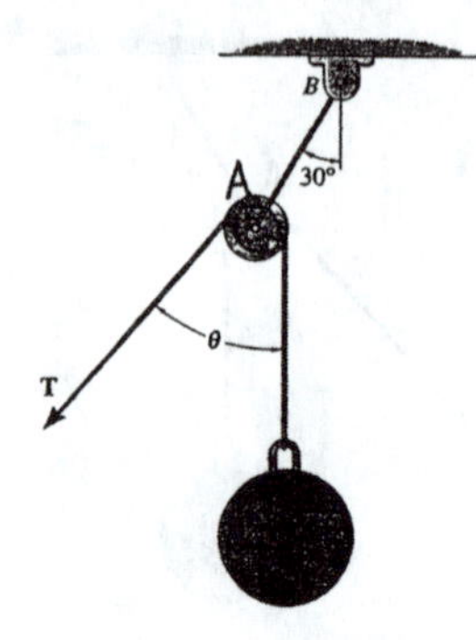

Solution

1. The pulley has *negligible shape* (we are told it is *small*) so that it can be modeled as a particle.
2. Imagine the pulley to be separated or detached from the system.
3. The (detached) pulley is subjected to three *external* forces. They are caused by:
   - **i. CORD** AB
   - **ii. FORCE T**
   - **iii. WEIGHT OF BALL**
4. Draw the free-body diagram of the (detached) pulley showing all these forces labeled with their magnitudes and directions. You should also include any other available information e.g. lengths, angles etc—which will help when formulating the equilibrium equations for the pulley.

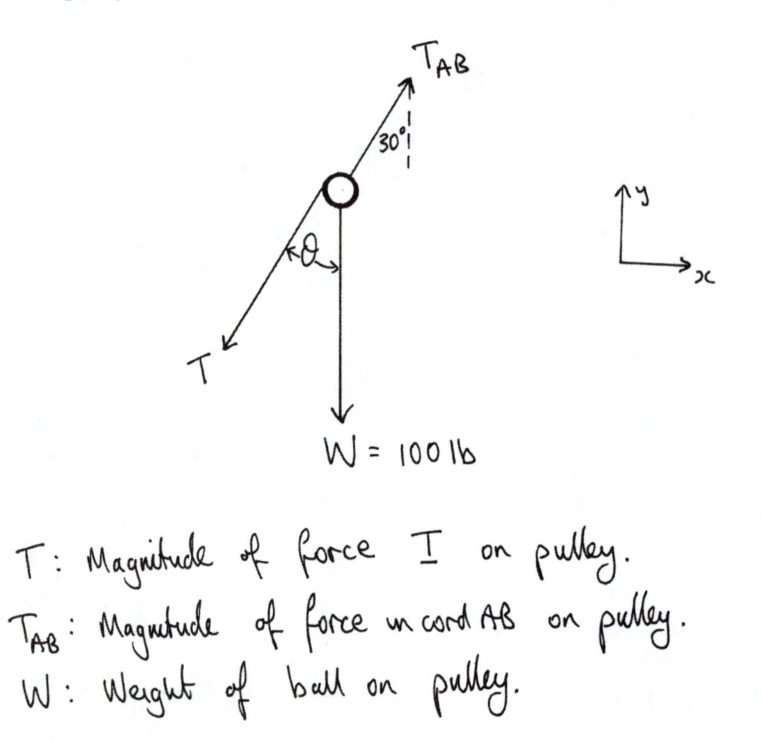

Problem 3.7.

The system is in equilibrium. Draw free-body diagrams for the block A and disk B.

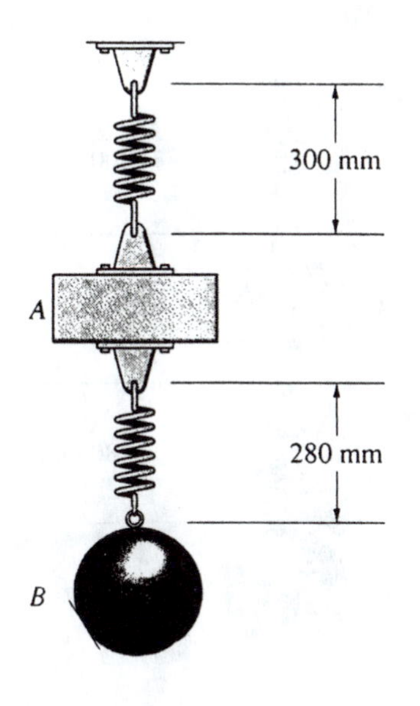

Solution

1. Imagine the block and the disk to be separated or detached from the system.
2. Block A is subjected to three *external* forces. They are caused by:

   i. **ii.**

   iii.

   Disk B is subjected to two external forces caused by:

   i. **ii.**

3. Draw the free-body diagrams of block A and disk B showing all these forces labeled with their magnitudes and directions.

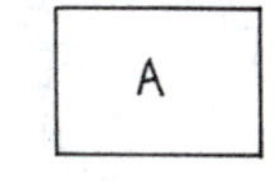

B

Problem 3.7.

The system is in equilibrium. Draw free-body diagrams for the block A and disk B.

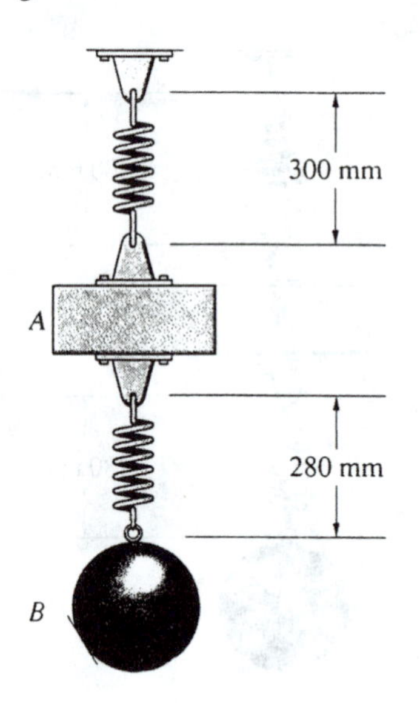

Solution

1. Imagine the block and the disk to be separated or detached from the system.
2. Block A is subjected to three *external* forces. They are caused by:

   i. TENSION IN UPPER SPRING **ii. TENSION IN LOWER SPRING**

   iii. WEIGHT OF BLOCK ***A***

   Disk B is subjected to two external forces caused by:

   i. TENSION IN LOWER SPRING **ii. WEIGHT OF DISK** ***B***
3. Draw the free-body diagrams of block A and disk B showing all these forces labeled with their magnitudes and directions.

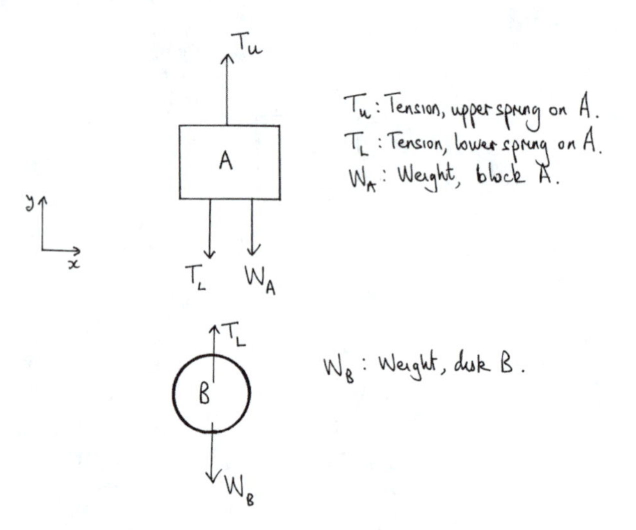

Problem 3.8.

The 50-kg load is supported at A by a system of five cords. Draw the free-body diagrams for the rings at A and B when the system is in equilibrium.

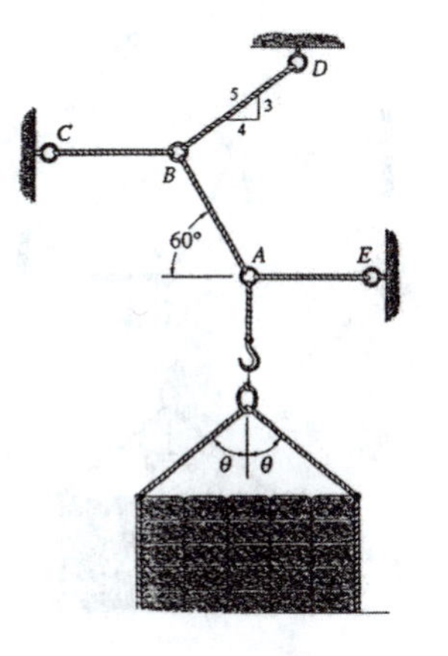

Solution

1. Imagine A and B to be separated or detached from the system.
2. Each of A and B is subjected to three *external* forces. For A, they are caused by:

   i. **ii.**

   iii.

   For B, they are caused by:

   i. **ii.**

   iii.
3. Draw the free-body diagrams of A and B showing all these forces labeled with their magnitudes and directions. You should also include any other available information e.g. lengths, angles etc—which will help when formulating the equilibrium equations.

Problem 3.8.

The 50-kg load is supported at A by a system of five cords. Draw the free-body diagrams for the rings at A and B when the system is in equilibrium.

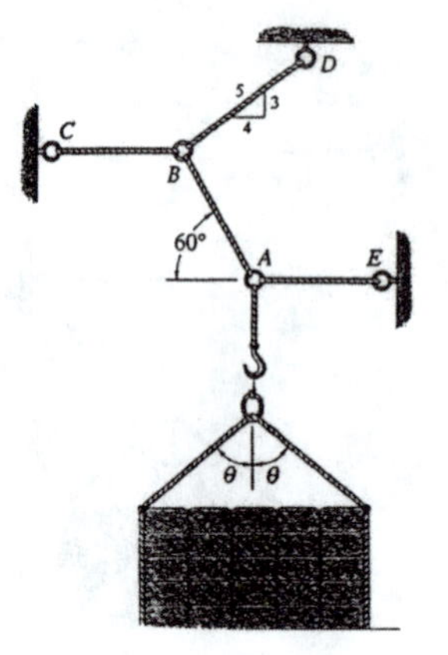

Solution

1. Imagine A and B to be separated or detached from the system.
2. Each of A and B is subjected to three *external* forces. For A, they are caused by:
   - **i. CABLE** AB
   - **ii. CABLE** AE
   - **iii. WEIGHT OF LOAD**

   For B, they are caused by:
   - **i. CABLE** BC
   - **ii. CABLE** BD
   - **iii. CABLE** BA
3. Draw the free-body diagrams of A and B showing all these forces labeled with their magnitudes and directions. You should also include any other available information e.g. lengths, angles etc—which will help when formulating the equilibrium equations.

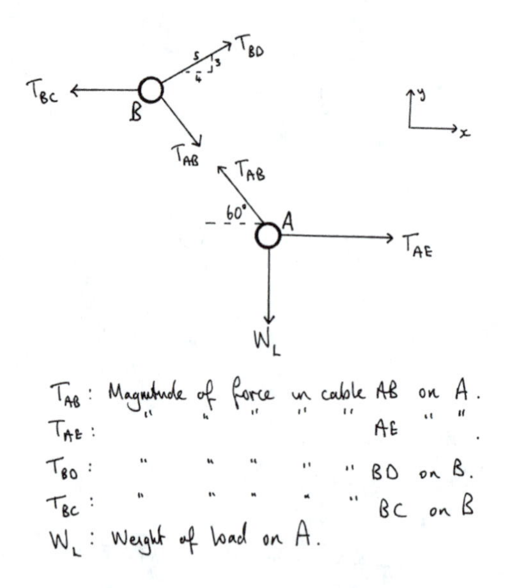

Problem 3.9.

The system is in equilibrium. Draw free-body diagrams for each pulley and the suspended object A.

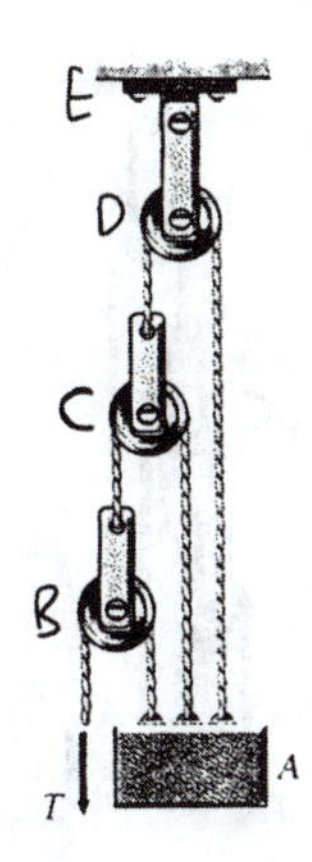

Solution

1. Imagine each pulley and the object to be separated or detached from the system.
2. Each (detached) pulley and the object at A is subjected to four *external* forces, *not* all independent.
3. Draw free-body diagrams of each (detached) pulley and the object showing all these forces labeled with their magnitudes and directions.

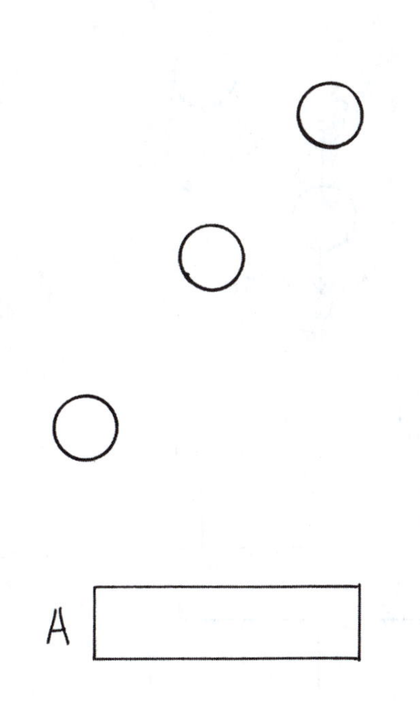

Problem 3.9.

The system is in equilibrium. Draw free-body diagrams for each pulley and the suspended object A.

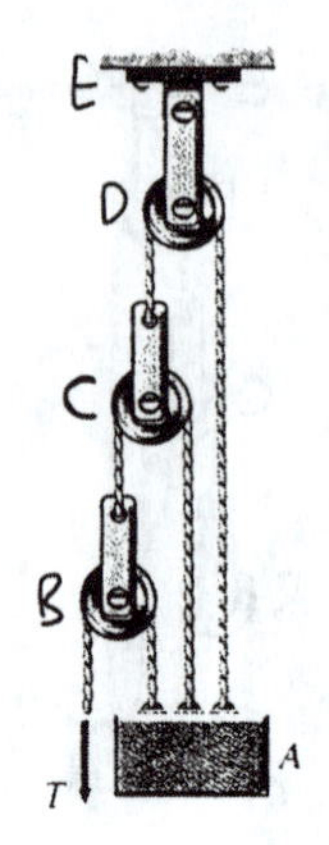

Solution

1. Imagine each pulley and the object to be separated or detached from the system.
2. Each (detached) pulley and the object at A is subjected to four *external* forces, *not* all independent.
3. Draw free-body diagrams of each (detached) pulley and the object showing all these forces labeled with their magnitudes and directions.

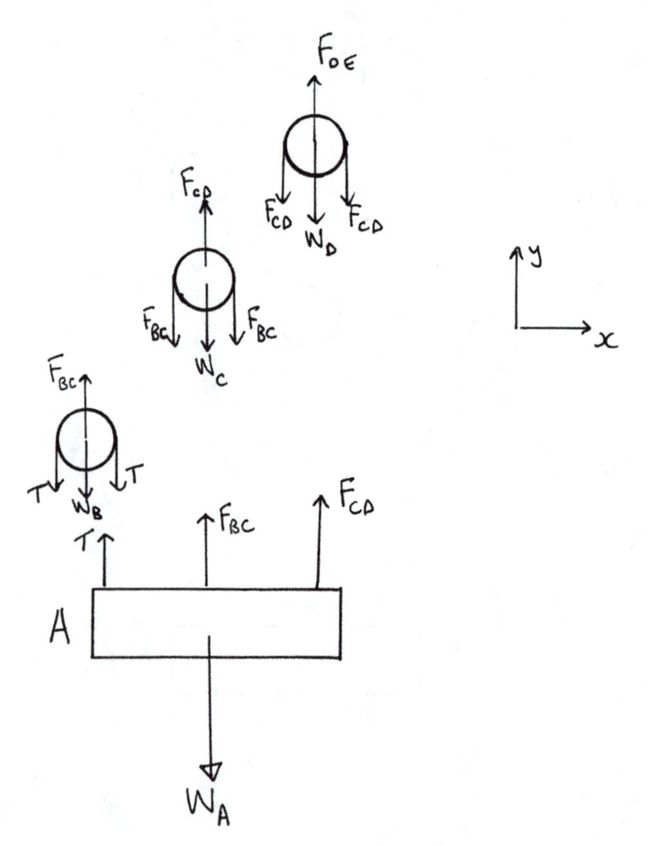

Problem 3.10.

The airplane is in steady flight. Draw a free-body diagram of the airplane and use it to write down the equilibrium equations for the airplane.

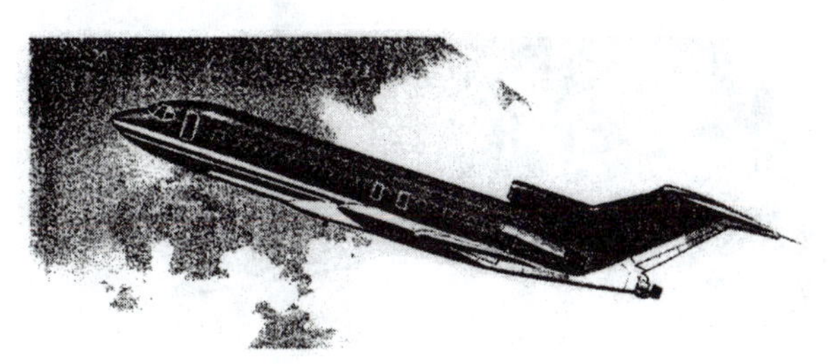

Solution

1. Imagine the airplane to be separated or detached from the system (airplane + sky).
2. The (detached) airplane is subjected to four *external* forces. They are caused by:

   i. **ii.**

   iii. **iv.**
3. Draw the free-body diagram of the (detached) airplane showing all these forces labeled with their magnitudes and directions. Include any other relevant information e.g. lengths, angles etc

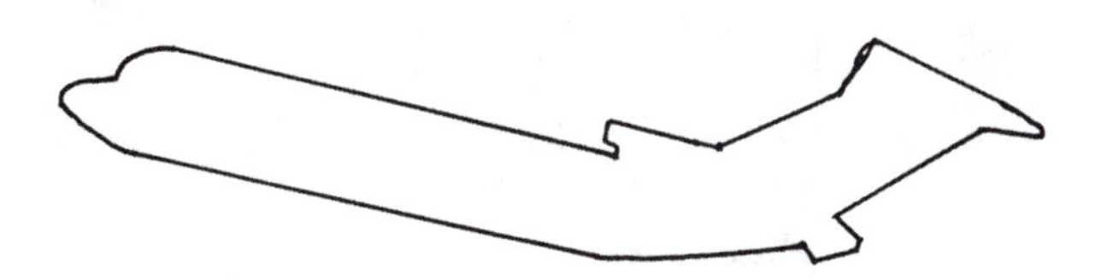

4. Establish an xy-axes system on the free-body diagram and write down the equilibrium equations in each of the x and $y-$ directions. Take the $x-axis$ to be along the path of the airplane.

$+\nwarrow \sum F_x = 0:$

$+\nearrow \sum F_y = 0:$

Problem 3.10.

The airplane is in steady flight. Draw a free-body diagram of the airplane and use it to write down the equilibrium equations for the airplane.

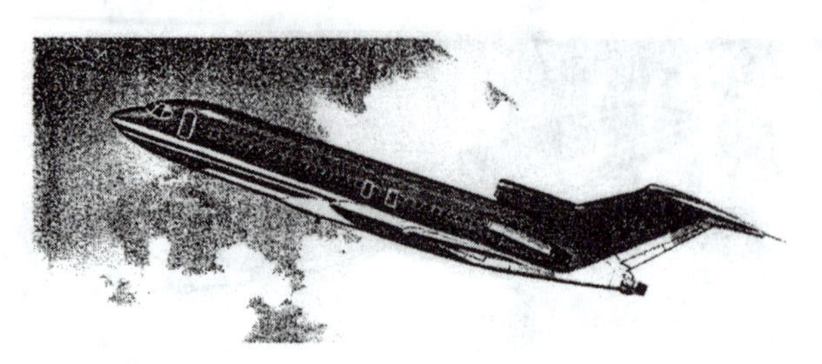

Solution

1. Imagine the airplane to be separated or detached from the system (airplane + sky).
2. The (detached) airplane is subjected to four *external* forces. They are caused by:

   i. DRAG D **ii. LIFT** L

   iii. THRUST T **iv. WEIGHT** W

3. Draw the free-body diagram of the (detached) airplane showing all these forces labeled with their magnitudes and directions. Include any other relevant information e.g. lengths, angles etc

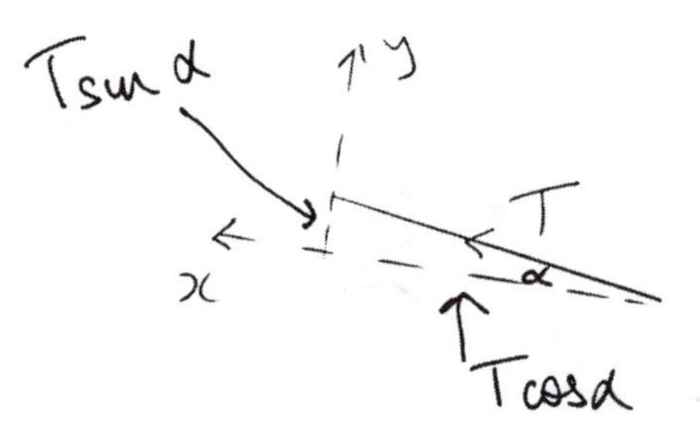

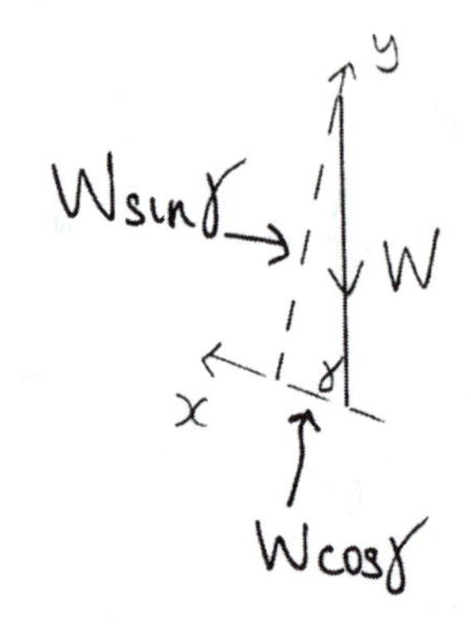

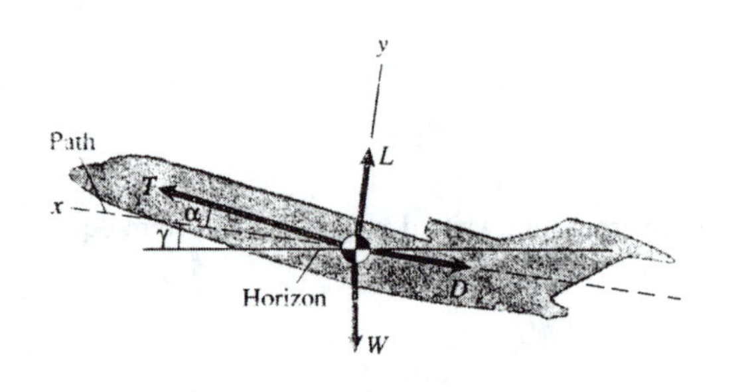

4. Establish an xy-axes system on the free-body diagram and write down the equilibrium equations in each of the x and $y-$ directions. Take the $x-axis$ to be along the path of the airplane.

$$+\nwarrow \sum F_x = 0: \quad T\cos\alpha - D - W\sin\gamma = 0$$

$$+\nearrow \sum F_y = 0: \quad T\sin\alpha + L - W\cos\gamma = 0$$

Problem 3.11.

The cord suspends the *small* bucket in the equilibrium position shown. The spring has an unstretched length of 3−ft and the system is in equilibrium at angle θ. Draw the free-body diagram of the connecting knot at A and write down the equilibrium equations for the knot at A.

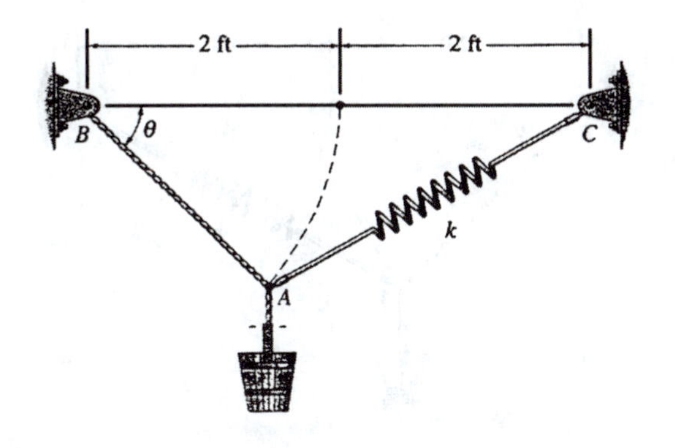

Solution

1. The knot at A has *negligible shape* so that it can be modelled as a particle.
2. Imagine the knot at A to be separated or detached from the system.
3. The (detached) knot at A is subjected to three *external* forces. They are caused by:

   i. **ii.**

   iii.

4. Draw the free-body diagram of the (detached) knot showing all these forces labeled with their magnitudes and directions.

A

5. Establish an xy-axes system on the free-body diagram and write down the equilibrium equations in each of the x and $y-$ directions

   $+\rightarrow \sum F_x = 0:$

   $+\uparrow \ \sum F_y = 0:$

Problem 3.11.

The cord suspends the *small* bucket in the equilibrium position shown. The spring has an unstretched length of 3−ft and the system is in equilibrium at angle θ. Draw the free-body diagram of the connecting knot at A and write down the equilibrium equations for the knot at A.

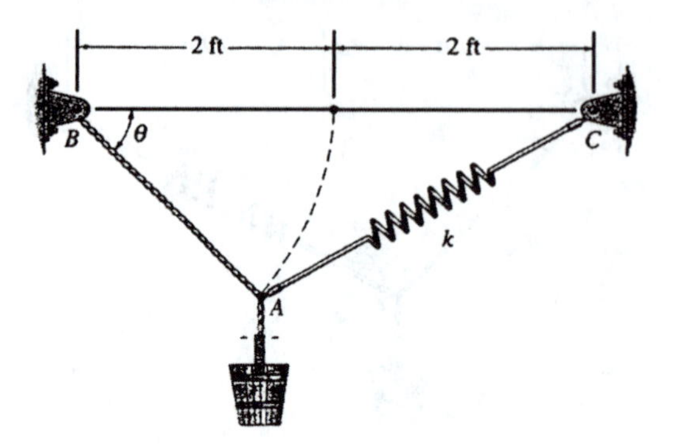

Solution

1. The knot at A has *negligible shape* so that it can be modelled as a particle.
2. Imagine the knot at A to be separated or detached from the system.
3. The (detached) knot at A is subjected to three *external* forces. They are caused by:
   i. **CORD** AB
   ii. **SPRING** AC
   iii. **WEIGHT OF BUCKET**
4. Draw the free-body diagram of the (detached) knot showing all these forces labeled with their magnitudes and directions.

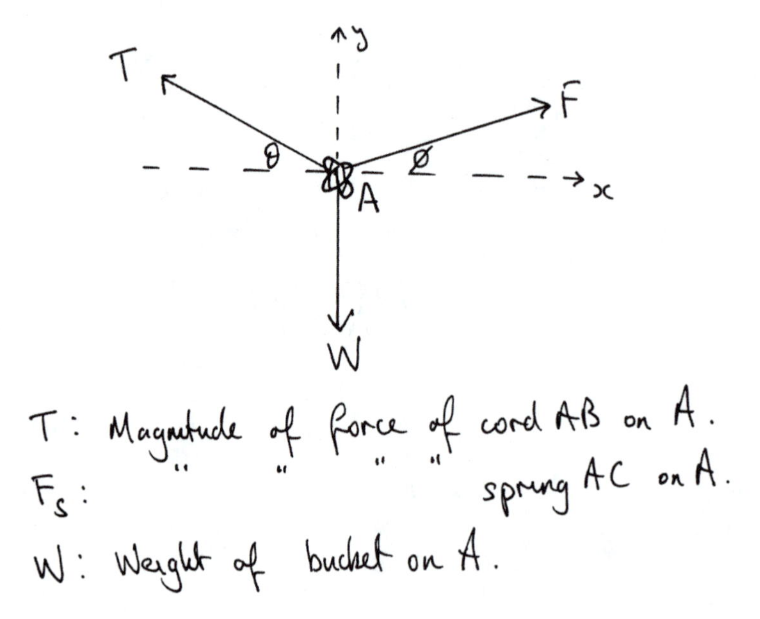

5. Establish an xy-axes system on the free-body diagram and write down the equilibrium equations in each of the x and y− directions

$$+\rightarrow \sum F_x = 0: \quad F_s \cos\phi - T\cos\theta = 0.$$

$$+\uparrow \sum F_y = 0: \quad T\sin\theta + F_s \sin\phi - W = 0.$$

Problem 3.12.

The mass of the cylinder A is 30 kg. Draw separate free-body diagrams for the cylinder A and the pulley B. Use these free-body diagrams to formulate equilibrium equations for both the cylinder and the pulley. Use these equilibrium equations to find the mass of the pulley B.

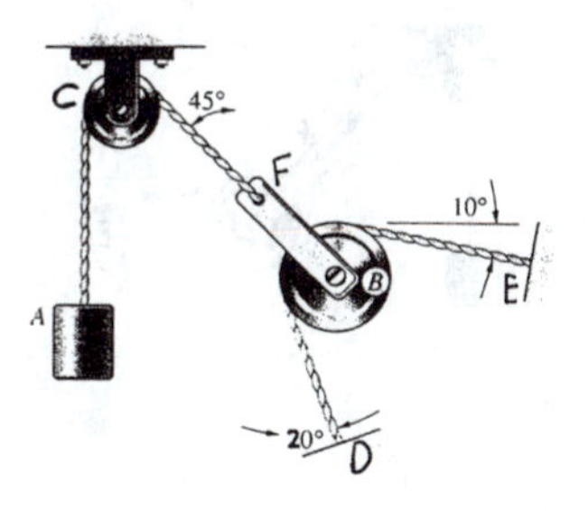

Solution

1. Imagine the cylinder A to be separated or detached from the system.
2. The (detached) cylinder is subjected to two *external* forces. They are caused by:

   i. **ii.**

3. Imagine the pulley B to be separated or detached from the system.
4. The (detached) pulley is subjected to four *external* forces.
   They are caused by:

   i. **ii.**

   iii. **iv.**

   (Note that the tensions in both sides of the cord going around pulley B have the same magnitude)
5. Draw the free-body diagram of the (detached) pulley showing these forces labeled with their magnitudes and directions.

A B

6. Establish an xy-axes system on each free-body diagram and write down the equilibrium equations for each of the cylinder A and the pulley B :

   Cylinder:
   $+\uparrow \ \sum F_y = 0:$
   Pulley:
   $+\rightarrow \sum F_x = 0:$

   $+\uparrow \ \sum F_y = 0:$

   You should have obtained three equations for three unknowns (the tensions in both sides of the cord going around pulley B have the same magnitude), one of which is m_{pulley} , the required mass of the pulley B. Solve these equations to obtain m_{pulley}.

Problem 3.12.

The mass of the cylinder A is 30 kg. Draw separate free-body diagrams for the cylinder A and the pulley B. Use these free-body diagrams to formulate equilibrium equations for both the cylinder and the pulley. Use these equilibrium equations to find the mass of the pulley B.

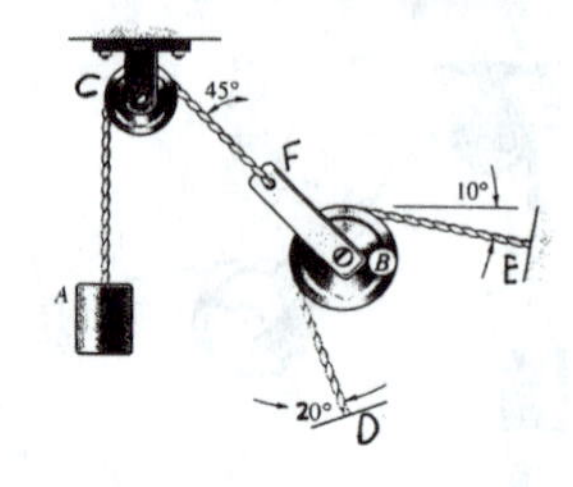

Solution

1. Imagine the cylinder A to be separated or detached from the system.
2. The (detached) cylinder is subjected to two *external* forces. They are caused by:

   i. TENSION IN CORD AC **ii. WEIGHT OF CYLINDER** A
3. Imagine the pulley B to be separated or detached from the system.
4. The (detached) pulley is subjected to four *external* forces. They are caused by:

   i. PULLEY'S WEIGHT **ii. TENSION IN CORD** BD

   iii. TENSION IN CORD BE **iv. FORCE ON PULLEY FROM** BF

   (Note that the tensions in both sides of the cord going around pulley B have the same magnitude)
5. Draw the free-body diagram of the (detached) pulley showing these forces labeled with their magnitudes and directions.

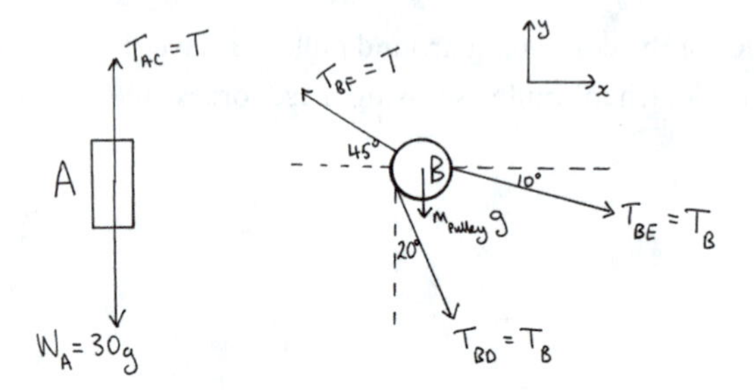

Frictionless pulleys at B and C: $T_{BE} = T_{BD} = T_B$; $T_{AC} = T_{BF} = T$.

6. Establish an xy-axes system on each free-body diagram and write down the equilibrium equations for each of the cylinder A and the pulley B :

   Cylinder:

   $+\uparrow \ \sum F_y = 0: \quad T - 30g = 0$

   Pulley:

   $+\rightarrow \ \sum F_x = 0: \quad -T\cos 45^\circ + T_B \cos 10^\circ + T_B \sin 20^\circ = 0$

   $+\uparrow \ \sum F_y = 0: \quad T\sin 45^\circ - T_B \sin 10^\circ - T_B \cos 20^\circ - m_{pulley} g = 0$ (Here, $T_B = T_{BE} = T_{BD}$ (frictionless pulley))

   You should have obtained three equations in three unknowns, one of which is m_{pulley} , the required mass of the pulley B. Solve these equations to obtain m_{pulley} :

   $T = 294.3N$

   $T_B = T_{BE} = T_{BD} = 156.8N$

   $m_{pulley} = 3.41kg$

Problem 3.13.

The post anchors a cable that helps support an oil derrick. If $\alpha = 35^\circ$ and $\beta = 50^\circ$, draw a free-body diagram for the ring (cable juncture) at A. Use this free-body diagram to formulate appropriate equilibrium equations. Use these equations to find the tensions in cables AB and AC in terms of the tension T.

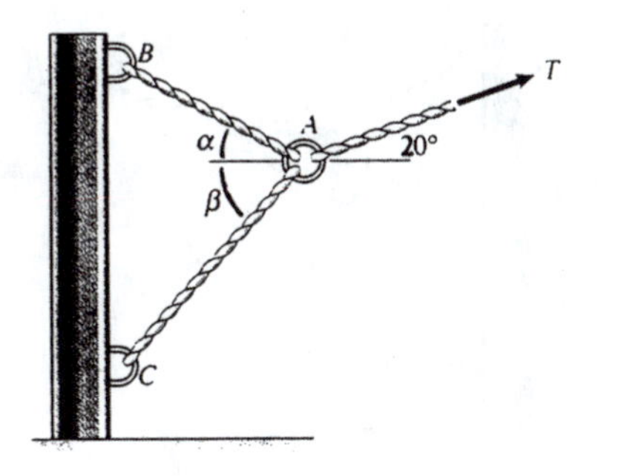

Solution

1. Imagine the ring at A to be separated or detached from the system.
2. The (detached) ring at A is subjected to three *external* forces. They are caused by:

   i. **ii.**

   iii.
3. Draw the free-body diagram of the (detached) ring showing all these forces labeled with their magnitudes and directions. Include also any other information which may help when formulating the equilibrium equations for the ring.

A

4. Establish an xy-axes system on the free-body diagram and write down the equilibrium equations in the x and $y-$ directions:

   $+\rightarrow \sum F_x = 0$

   $+\uparrow \ \sum F_y = 0:$
5. Solve for the required tensions in terms of the tension T:

Problem 3.13.

The post anchors a cable that helps support an oil derrick. If $\alpha = 35^\circ$ and $\beta = 50^\circ$, draw a free-body diagram for the ring (cable juncture) at A. Use this free-body diagram to formulate appropriate equilibrium equations. Use these equations to find the tensions in cables AB and AC in terms of the tension T.

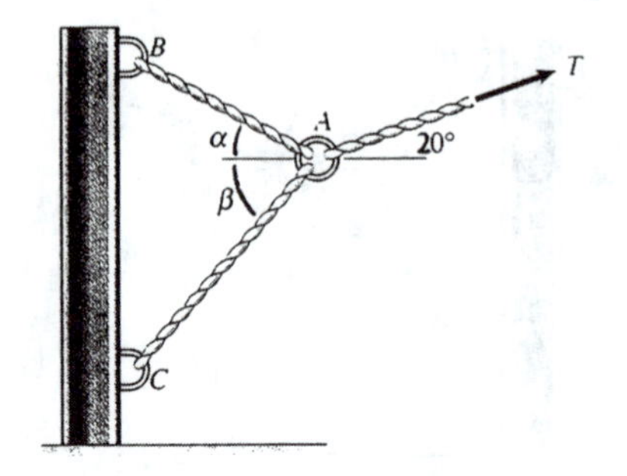

Solution

1. Imagine the ring at A to be separated or detached from the system.
2. The (detached) ring at A is subjected to three *external* forces. They are caused by:

   i. CABLE AB **ii. CABLE** AC

   iii. FORCE OF MAGNITUDE T
3. Draw the free-body diagram of the (detached) ring showing all these forces labeled with their magnitudes and directions. Include also any other information which may help when formulating the equilibrium equations for the ring.

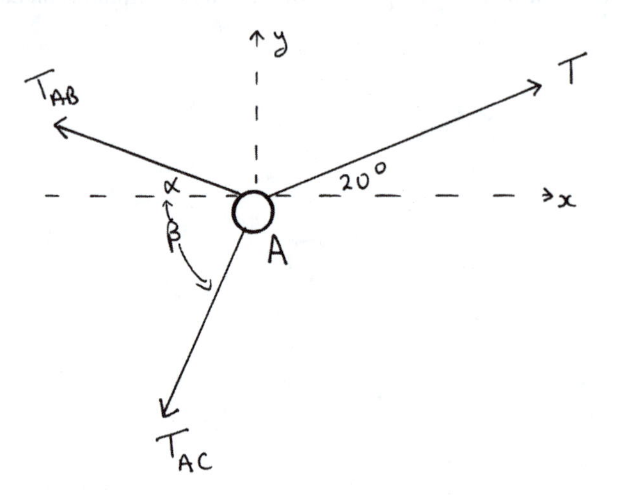

4. Establish an xy-axes system on the free-body diagram and write down the equilibrium equations in the x and $y-$ directions:

$$+\rightarrow \sum F_x = 0: \quad T\cos 20^\circ - T_{AB}\cos\alpha - T_{AC}\cos\beta = 0$$

$$+\uparrow \sum F_y = 0: \quad T\sin 20^\circ + T_{AB}\sin\alpha - T_{AC}\sin\beta = 0$$

5. Solve for the required tensions in terms of the tension T, noting that $\alpha = 35^\circ$ and $\beta = 50^\circ$:

$$T_{AC} = \frac{\sin(\alpha + 20^\circ)}{\sin(\alpha + \beta)} T = 0.8223T$$

$$T_{AB} = \frac{\sin(\beta - 20^\circ)}{\sin(\alpha + \beta)} T = 0.5019T$$

Problem 3.14.

The following system lies on the surface of the planet Mars (acceleration due to gravity $= 4.02m/s^2 \downarrow$). The unstretched length of the spring AB is 660 mm and the spring constant $k = 1000$ N/m. draw a free-body diagram of the connector at A and use it to find the mass of the object suspended from A.

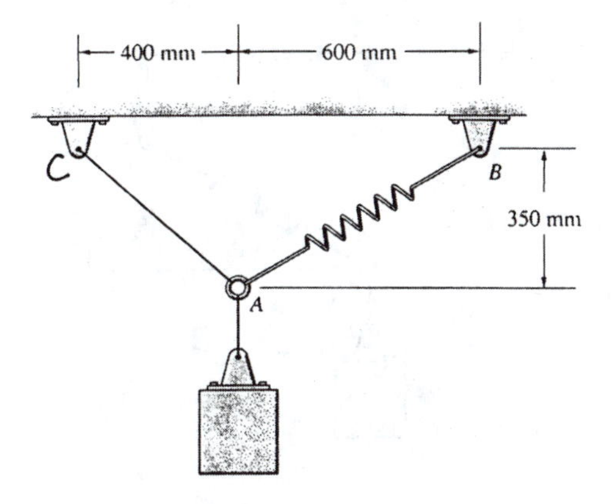

Solution

1. Imagine the connector to be separated or detached from the system.
2. The (detached) connector is subjected to three *external* forces. They are caused by:

   i. **ii.**

   iii.
3. Draw the free-body diagram of the (detached) connector showing all these forces labeled with their magnitudes and directions. Include any other relevant information e.g. lengths, angles etc

A

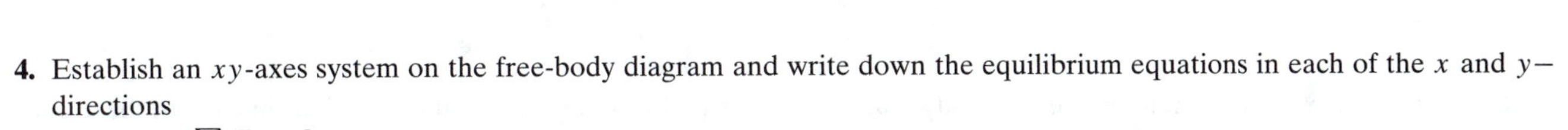

4. Establish an xy-axes system on the free-body diagram and write down the equilibrium equations in each of the x and $y-$ directions

   $+\rightarrow \sum F_x = 0:$

   $+\uparrow \ \sum F_y = 0:$
5. Find the extension of and hence the magnitude of the tension in the spring AB using the linear spring force-extension relation:
6. Solve the above three equations for three unknowns including the weight of the suspended object:
7. Find the mass of the suspended object:

Problem 3.14.

The following system lies on the surface of the planet Mars (acceleration due to gravity $= 4.02 m/s^2 \downarrow$). The unstretched length of the spring AB is 660 mm and the spring constant $k = 1000$ N/m. Draw a free-body diagram of the connector at A and use it to find the mass of the object suspended from A.

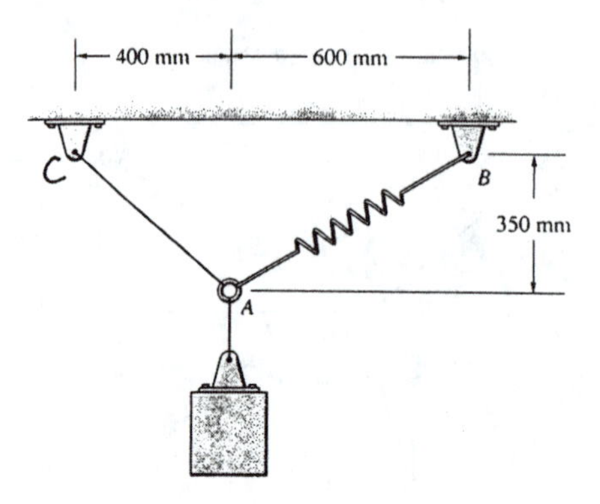

Solution

1. Imagine the connector to be separated or detached from the system.
2. The (detached) connector is subjected to three *external* forces. They are caused by:
   i. **CABLE** AC
   ii. **SPRING** AB
   iii. **WEIGHT OF OBJECT AT** A
3. Draw the free-body diagram of the (detached) connector showing all these forces labeled with their magnitudes and directions. Include any other relevant information e.g. lengths, angles etc

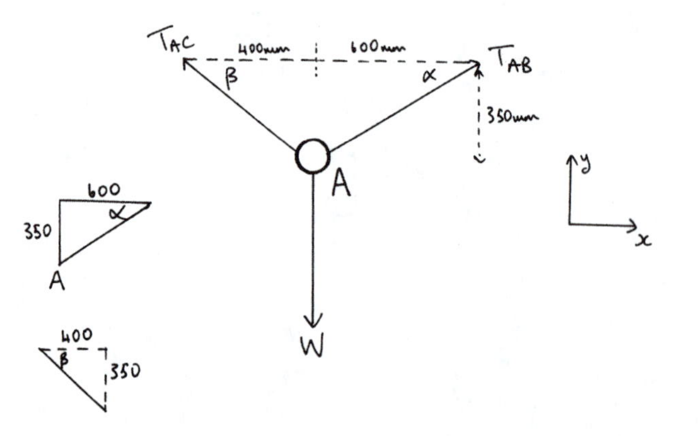

4. Establish an xy-axes system on the free-body diagram and write down the equilibrium equations in each of the x and $y-$ directions

$$+\rightarrow \sum F_x = 0: \quad T_{AB}\cos\alpha - T_{AC}\cos\beta = 0 \qquad (1)$$

$$+\uparrow \sum F_y = 0: \quad T_{AC}\sin\alpha - T_{AB}\sin\beta - W = 0 \qquad (2)$$

5. Find the extension of and hence the magnitude of the tension in the spring AB using the linear spring force-extension relation:Spring extension is: $\Delta L = \sqrt{(350)^2 + (600)^2} - 660 = 34.62mm$ $T_{AB} = k\Delta L = (1000)\,(0.03462) = 34.6N$ (3)
6. Solve the above three equations for three unknowns including the weight of the suspended object: Noting that $\tan\alpha = \frac{350}{600}$, $\tan\beta = \frac{350}{400}$, we obtain: $\alpha = 30.26°$, $\beta = 41.1°$ Solving the three equations (1) - (3) leads to: $W = 43.62N$
7. Find the mass of the suspended object: $m = \frac{W}{g} = \frac{43.62}{4.02} = 10.85kg$.

Problem 3.15.

The breeches buoy is used to transfer the person B between two ships. The person is attached to a pulley that rolls on the overhead cable. The total weight of the person and the buoy is 250 lb. Draw a free-body diagram of the person at B and use it to find the tension in the horizontal line AB necessary to hold the person in equilibrium in the position shown.

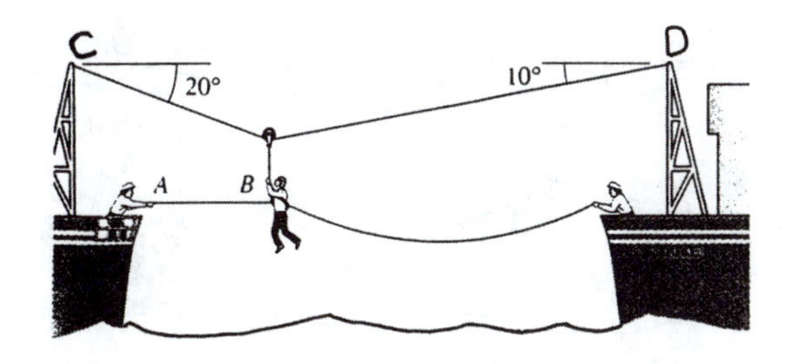

Solution

1. The person at B has *negligible shape* and so can be modelled as a particle.
2. Imagine the person to be separated or detached from the system.
3. The (detached) person is subjected to four *external* forces. They are caused by:

   i. **ii.**

   iii. **iv.**

4. Draw the free-body diagram of the (detached) person showing all these forces labeled with their magnitudes and directions. Include any other relevant information e.g. lengths, angles etc

5. Establish an xy-axes system on the free-body diagram and write down the equilibrium equations in each of the x and $y-$ directions

   $+\rightarrow \sum F_x = 0:$

   $+\uparrow \ \sum F_y = 0:$

6. Solve these equations to find the required tension in the line AB :

Problem 3.15.

The breeches buoy is used to transfer the person B between two ships. The person is attached to a pulley that rolls on the overhead cable. The total weight of the person and the buoy is 250 lb. Draw a free-body diagram of the person at B and use it to find the tension in the horizontal line AB necessary to hold the person in equilibrium in the position shown.

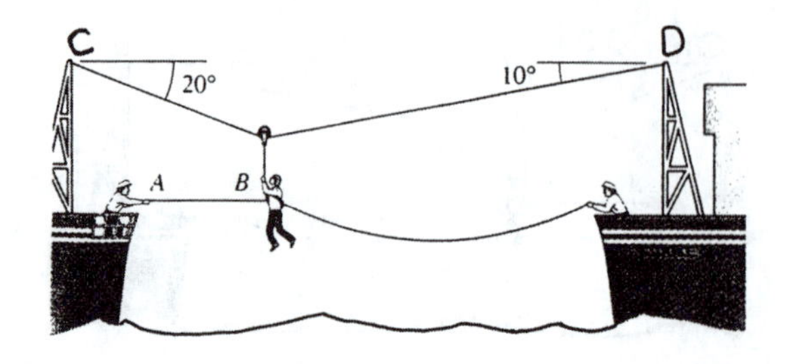

Solution

1. The person at B has *negligible shape* and so can be modelled as a particle.
2. Imagine the person to be separated or detached from the system.
3. The (detached) person is subjected to four *external* forces. They are caused by:
   - **i. CABLE** BC
   - **ii. CABLE** BD
   - **iii. CABLE** BA
   - **iv. TOTAL WEIGHT OF PERSON AND BUOY** ($W = 250$ lb)
4. Draw the free-body diagram of the (detached) person showing all these forces labeled with their magnitudes and directions. Include any other relevant information e.g. lengths, angles etc

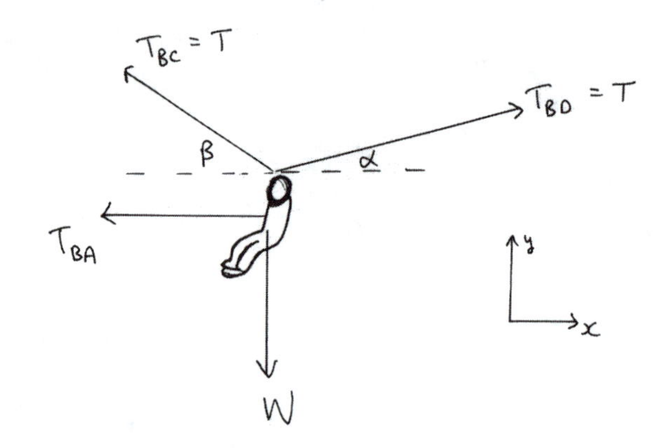

5. Establish an xy-axes system on the free-body diagram and write down the equilibrium equations in each of the x and $y-$ directions:

   Noting that $T_{BC} = T_{BD} = T$ (Frictionless pulley), $\alpha = 10^\circ$ and $\beta = 20^\circ$, we have that

   $+\!\rightarrow \sum F_x = 0: \quad -T\cos\beta + T\cos\alpha - T_{BA} = 0$

   $+\!\uparrow \sum F_y = 0: \quad T\sin\alpha + T\sin\beta - W = 0$

6. Solve these equations to find the required tension in the line AB :

   $$T_{BA} = \left(\frac{\cos\alpha - \cos\beta}{\sin\alpha + \sin\beta}\right) W = 21.87\ lb.$$

3.2 Free-Body Diagrams in the Equilibrium of a Rigid Body

In each of the following problems, assume that the objects are in equilibrium.

Problem 3.16.

Draw the free-body diagram of the 50-kg uniform bar supported at A and B.

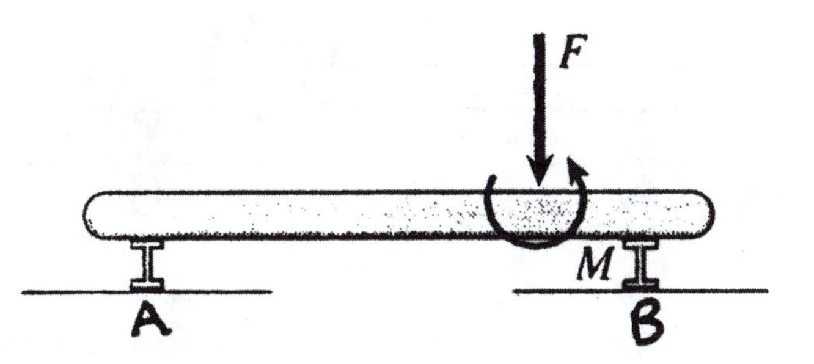

Solution

1. Imagine the bar to be separated or detached from the system.
2. The supports at A and B are equivalent to Roller Supports. Use Table 2.1 to determine the number and types of reactions *acting on the bar* at A and B.
3. The bar is subjected to four *external* forces (don't forget the weight!). They are caused by:

   i. **ii.**

   iii. **iv.**

   The bar is subjected also to one applied external couple moment with magnitude M.
4. Draw the free-body diagram of the (detached) bar showing all these forces and any couples labeled with their magnitudes and directions. *Assume* the sense of the vectors representing the *reactions acting on the bar* (the correct sense will always emerge from the equilibrium equations for the bar). Include any other relevant information e.g. lengths, angles etc which may help when formulating the equilibrium equations (including the moment equation) for the pipe.

Problem 3.16.

Draw the free-body diagram of the 50-kg uniform bar supported at A and B.

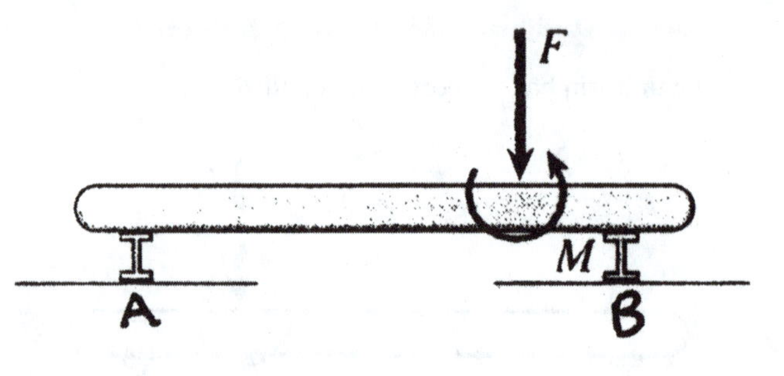

Solution

1. Imagine the bar to be separated or detached from the system.
2. The supports at A and B are equivalent to Roller Supports. Use Table 2.1 to determine the number and types of reactions *acting on the bar* at A and B.
3. The bar is subjected to four *external* forces (don't forget the weight!). They are caused by:

   i. REACTION AT A **ii. REACTION AT** B

   iii. APPLIED FORCE F **iv. WEIGHT OF BAR**

   The bar is subjected also to one applied external couple with magnitude M.
4. Draw the free-body diagram of the (detached) bar showing all these forces and any couples labeled with their magnitudes and directions. *Assume* the sense of the vectors representing the *reactions acting on the bar* (the correct sense will always emerge from the equilibrium equations for the bar). Include any other relevant information e.g. lengths, angles etc which may help when formulating the equilibrium equations (including the moment equation) for the pipe.

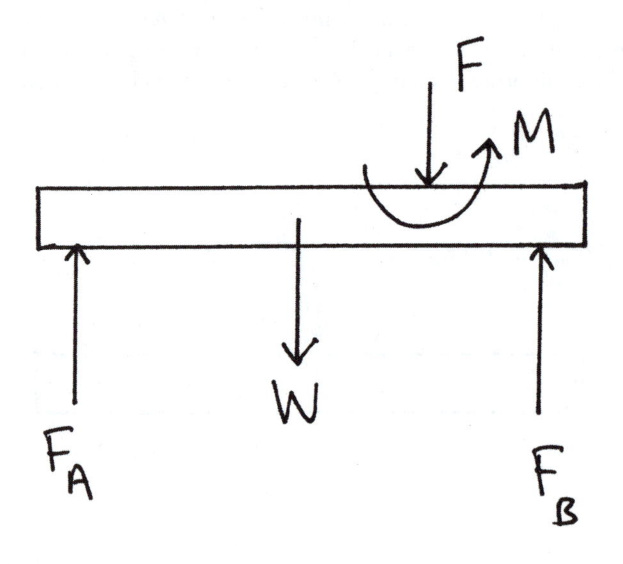

Problem 3.17.

Draw the free-body diagram of the bar which is built-in at A. Neglect the weight of the bar.

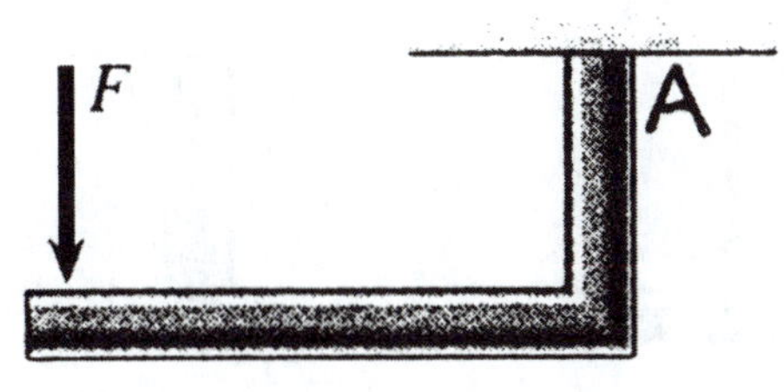

Solution

1. Imagine the bar to be separated or detached from the system.
2. The support at A is a built-in support. Use Table 2.1 to determine the number and types of reactions *acting on the bar* at A.
3. The bar is subjected to three *external* forces. They are caused by:

   i. **ii.**

   iii.

   The bar is subjected to one *external* couple. It is caused by:

4. Draw the free-body diagram of the (detached) bar showing all these forces and any couples labeled with their magnitudes and directions. *Assume* the sense of the vectors representing the *reactions acting on the bar* (the correct sense will always emerge from the equilibrium equations for the bar). Include any other relevant information e.g. lengths, angles etc which may help when formulating the equilibrium equations (including the moment equation) for the bar.

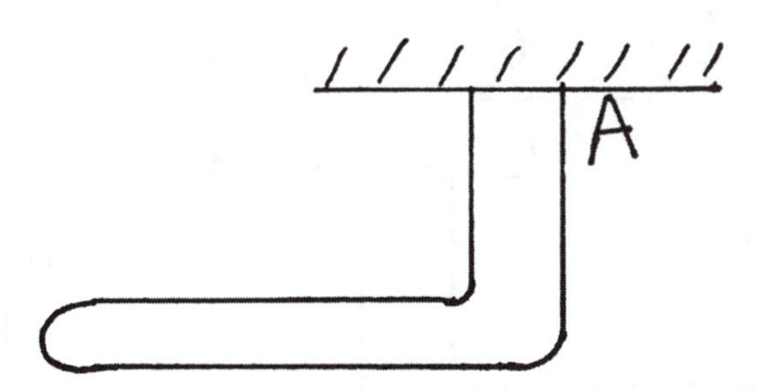

Problem 3.17.

Draw the free-body diagram of the bar which is built-in at A. Neglect the weight of the bar.

Solution

1. Imagine the bar to be separated or detached from the system.
2. The support at A is a built-in support. Use Table 2.1 to determine the number and types of reactions *acting on the bar* at A.
3. The bar is subjected to three *external* forces. They are caused by:

   i. REACTION AT A **(2 forces)** **ii. APPLIED FORCE F**

   The bar is subjected to one *external* couple. It is caused by:
   REACTION AT A
4. Draw the free-body diagram of the (detached) bar showing all these forces and any couples labeled with their magnitudes and directions. *Assume* the sense of the vectors representing the *reactions acting on the bar* (the correct sense will always emerge from the equilibrium equations for the bar). Include any other relevant information e.g. lengths, angles etc which may help when formulating the equilibrium equations (including the moment equation) for the bar.

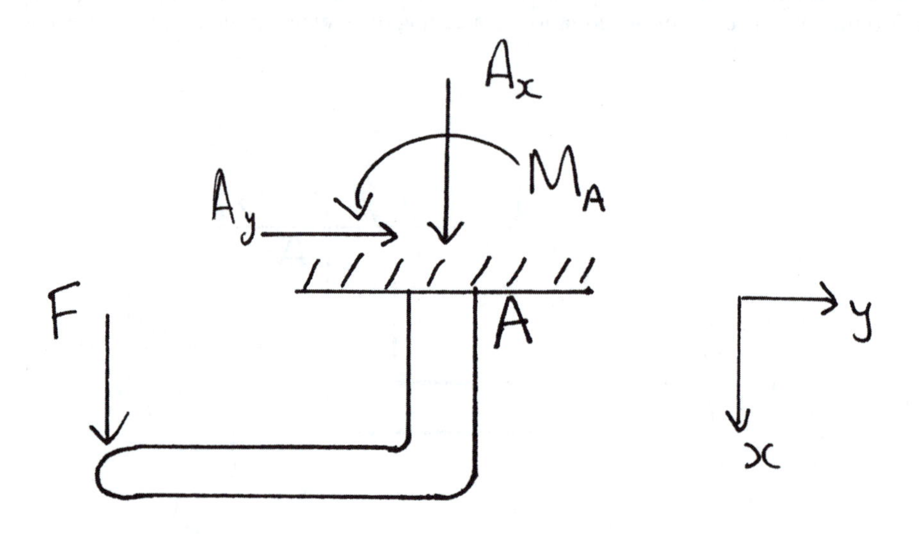

Problem 3.18.

Draw the free-body diagram of the beam which is supported by smooth surfaces at A and B and a rope at C. Neglect the weight of the beam.

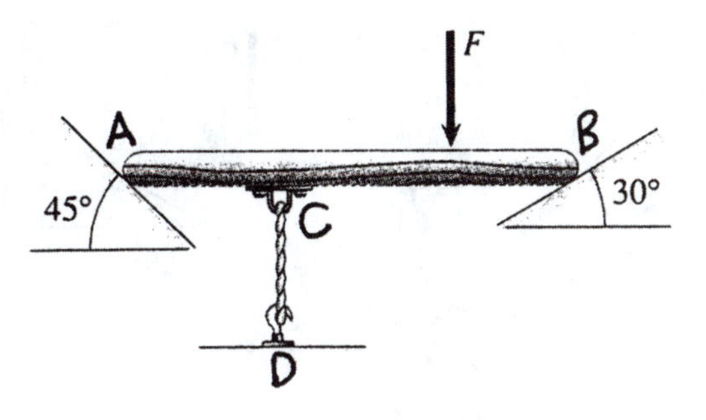

Solution

1. Imagine the beam to be separated or detached from the system.
2. The beam contacts the smooth surfaces at A and B and is supported additionally by a rope at C. Use Table 2.1 to determine the number and types of reactions *acting on the beam* at A, B and C.
3. The beam is subjected to four *external* forces. They are caused by:

   i. **ii.**

   iii. **iv.**

4. Draw the free-body diagram of the (detached) beam showing all these forces labeled with their magnitudes and directions. *Assume* the sense of the vectors representing the *reactions acting on the beam* (the correct sense will always emerge from the equilibrium equations for the beam). Include any other relevant information e.g. lengths, angles etc which may help when formulating the equilibrium equations (including the moment equation) for the beam.

Problem 3.18.

Draw the free-body diagram of the beam which is supported by smooth surfaces at A and B and a rope at C. Neglect the weight of the beam.

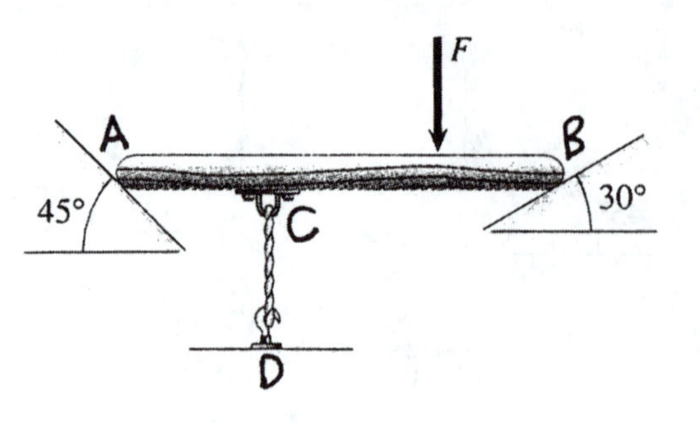

Solution

1. Imagine the beam to be separated or detached from the system.
2. The beam contacts the smooth surfaces at A and B and is supported additionally by a rope at C. Use Table 2.1 to determine the number and types of reactions *acting on the beam* at A, B and C.
3. The beam is subjected to four *external* forces. They are caused by:

   i. **REACTION AT** A
   ii. **REACTION AT** B
   iii. **APPLIED FORCE F**
   iv. **CORD** CD

4. Draw the free-body diagram of the (detached) beam showing all these forces labeled with their magnitudes and directions. *Assume* the sense of the vectors representing the *reactions acting on the beam* (the correct sense will always emerge from the equilibrium equations for the beam). Include any other relevant information e.g. lengths, angles etc which may help when formulating the equilibrium equations (including the moment equation) for the beam.

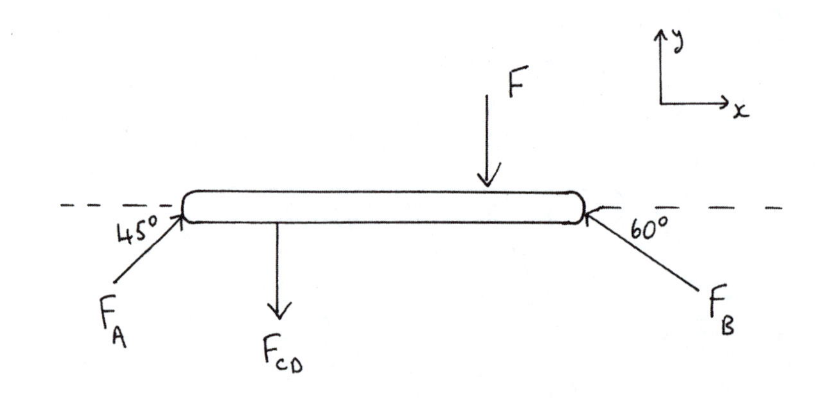

Problem 3.19.

Draw the free-body diagram of the plate which has a weight of 50 lb. The plate is supported by a pin connection at A and a constrained pin at B.

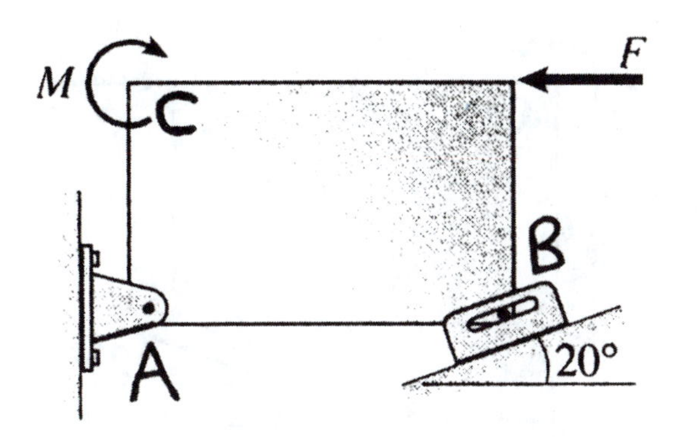

Solution

1. Imagine the plate to be separated or detached from its supports.
2. There is a pin support at A and and an inclined constrained pin support at B. Use Table 2.1 to determine the number and types of reactions *acting on the plate* at A and B.
3. The plate is subjected to five *external* forces. They are caused by:

   i. **ii.**

   iii. **iv.**

   v.

   In addition, the plate is subjected to one *external* applied couple of magnitude M at C.
4. Draw the free-body diagram of the (detached) plate showing all these forces and any couples labeled with their magnitudes and directions. *Assume* the sense of the vectors representing the *reactions acting on the plate* (the correct sense will always emerge from the equilibrium equations for the plate). Include any other relevant information e.g. lengths, angles etc which may help when formulating the equilibrium equations (including the moment equation) for the plate.

Problem 3.19.

Draw the free-body diagram of the plate which has a weight of 50 lb. The plate is supported by a pin connection at A and a constrained pin at B.

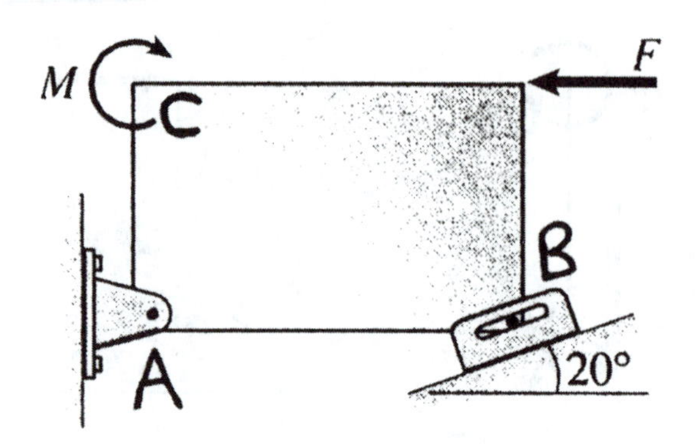

Solution

1. Imagine the plate to be separated or detached from its supports.
2. There is a pin support at A and and an inclined constrained pin support at B. Use Table 2.1 to determine the number and types of reactions *acting on the plate* at A and B.
3. The plate is subjected to five *external* forces.
   They are caused by:
   i. REACTION AT A **(2 forces)**　　**ii. REACTION AT** B
   iii. APPLIED FORCE F　　**iv. WEIGHT OF PLATE**
   In addition, the plate is subjected to one applied couple of magnitude M at C.
4. Draw the free-body diagram of the (detached) plate showing all these forces and any couples labeled with their magnitudes and directions. *Assume* the sense of the vectors representing the *reactions acting on the plate* (the correct sense will always emerge from the equilibrium equations for the plate). Include any other relevant information e.g. lengths, angles etc which may help when formulating the equilibrium equations (including the moment equation) for the plate.

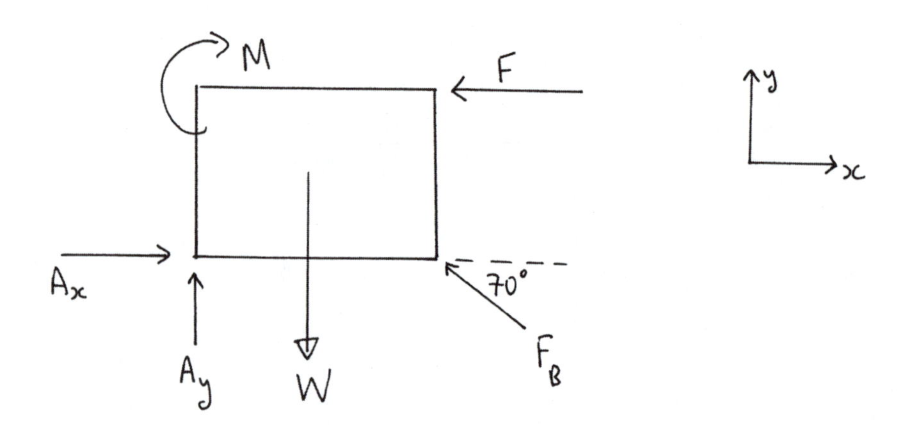

Problem 3.20.

Draw the free-body diagram of the link AC, which is (constrained-) pin-connected at A and rocker (roller) - supported at C. Neglect the weight of the link.

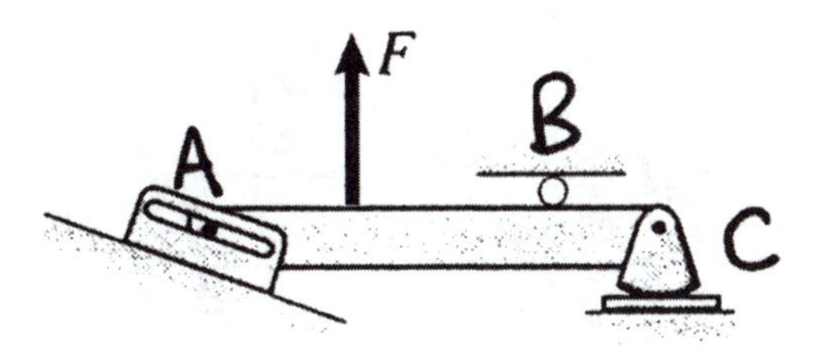

Solution

1. Imagine the link AC to be separated or detached from the system.
2. There is a pin connection at A, a rocker support at C and a roller support acts on the link at B. Use Table 2.1 to determine the number and types of reactions *acting on the link* at A, B and C.
3. The link is subjected to four *external* forces.
   They are caused by:

   i. **ii.**

   iii. **iv.**

4. Draw the free-body diagram of the (detached) link showing all these forces labeled with their magnitudes and directions. *Assume* the sense of the vectors representing the *reactions acting on the link* (the correct sense will always emerge from the equilibrium equations for the link). Include any other relevant information e.g. lengths, angles etc which may help when formulating the equilibrium equations (including the moment equation) for the link.

Problem 3.20.

Draw the free-body diagram of the link AC, which is (constrained-) pin-connected at A and rocker (roller) - supported at C. Neglect the weight of the link.

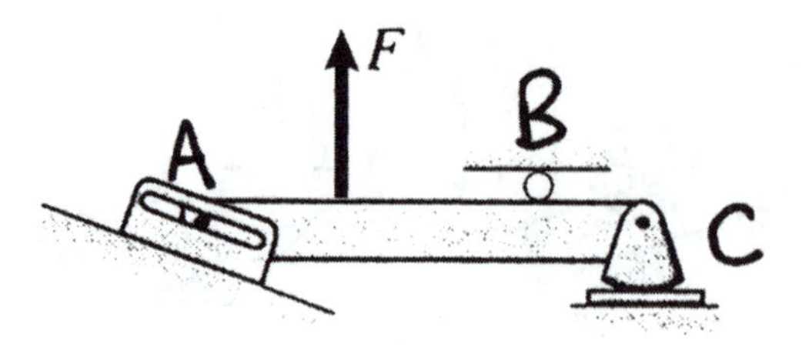

Solution

1. Imagine the link AC to be separated or detached from the system.
2. There is a pin connection at A, a rocker support at C and a roller support acts on the link at B. Use Table 2.1 to determine the number and types of reactions *acting on the link* at A, B and C.
3. The link is subjected to four *external* forces.
   They are caused by:

   i. **REACTION AT** A
   ii. **REACTION AT** B
   iii. **REACTION AT** C
   iv. **APPLIED FORCE F**

4. Draw the free-body diagram of the (detached) link showing all these forces labeled with their magnitudes and directions. *Assume* the sense of the vectors representing the *reactions acting on the link* (the correct sense will always emerge from the equilibrium equations for the link). Include any other relevant information e.g. lengths, angles etc which may help when formulating the equilibrium equations (including the moment equation) for the link.

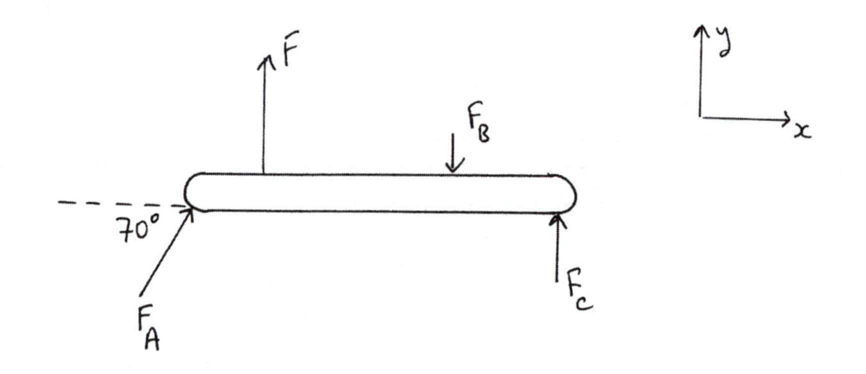

Problem 3.21.

Draw the free-body diagram of the uniform bar which has a mass of 10 kg and a center of mass at G. The bar contacts a rough surface at B.

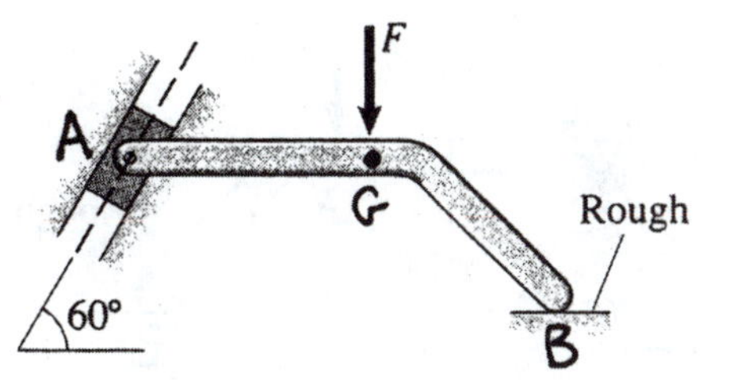

Solution

1. Imagine the bar to be separated or detached from the system.
2. The bar rests on a rough surface at B while end A is constrained by a slider. Use Table 2.1 to determine the number and types of reactions *acting on the bar* at A and B.
3. The bar is subjected to five *external* forces. They are caused by:

   i. **ii.**

   iii. **iv.**

   v.
4. Draw the free-body diagram of the (detached) bar showing all these forces labeled with their magnitudes and directions. *Assume* the sense of the vectors representing the *reactions acting on the bar* (the correct sense will always emerge from the equilibrium equations for the bar). Include any other relevant information e.g. lengths, angles etc which may help when formulating the equilibrium equations for the bar.

Problem 3.21.

Draw the free-body diagram of the uniform bar which has a mass of 10 kg and a center of mass at G. The bar contacts a rough surface at B.

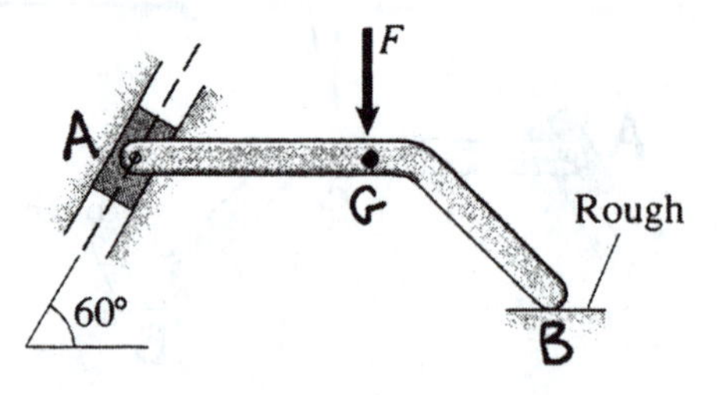

Solution

1. Imagine the bar to be separated or detached from the system.
2. The bar rests on a rough surface at B while end A is constrained by a slider. Use Table 2.1 to determine the number and types of reactions *acting on the bar* at A and B.
3. The bar is subjected to five *external* forces. They are caused by:

   i. **REACTION AT** B **(2 forces)**
   ii. **REACTION AT** A
   iii. **APPLIED FORCE F**
   iv. **WEIGHT OF BAR**

4. Draw the free-body diagram of the (detached) bar showing all these forces labeled with their magnitudes and directions. *Assume* the sense of the vectors representing the *reactions acting on the bar* (the correct sense will always emerge from the equilibrium equations for the bar). Include any other relevant information e.g. lengths, angles etc which may help when formulating the equilibrium equations for the bar.

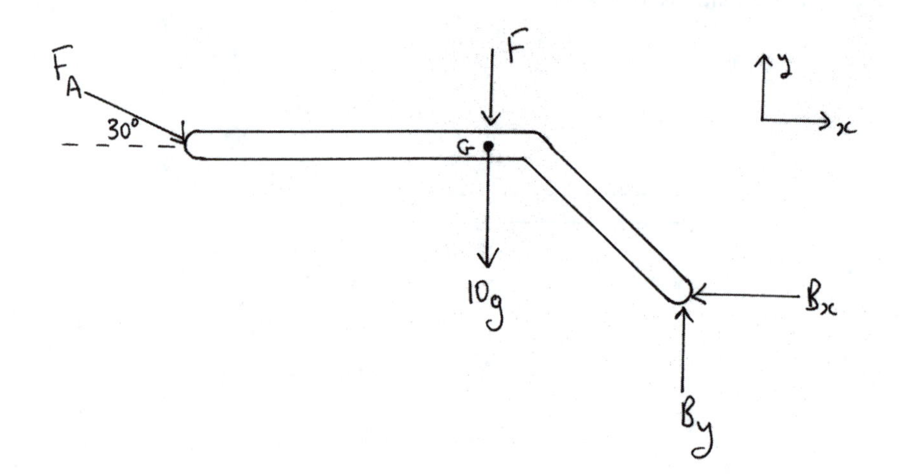

Problem 3.22.

In the following figure, the tension in cable AB is T. Draw a free-body diagram of the assembly and cable AB, treating them as a single object. Neglect the weight of the pipe assembly.

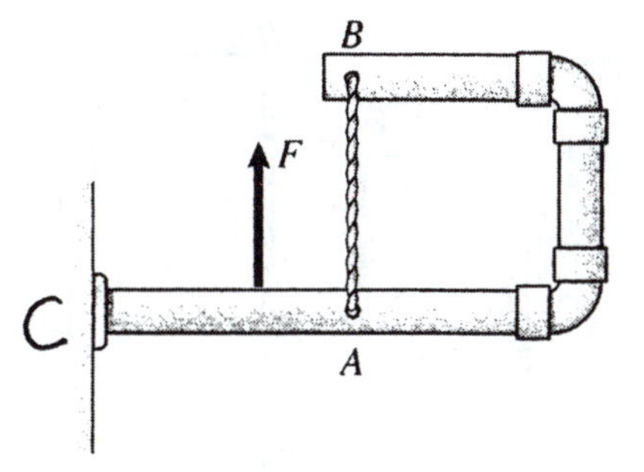

Solution

1. Imagine the entire pipe assembly to be separated or detached from the system (the supporting wall).
2. There is a built-in (fixed) support at C. Use Table 2.1 to determine the number and types of reactions *acting on the assembly* at C.
3. In addition to the force with magnitude F shown in the figure, the assembly is subjected to two additional *external* forces and an external couple. They are caused by:

   There are no other external forces acting on the assembly.
4. Draw the free-body diagram of the (detached) assembly showing all these forces and any couples labeled with their magnitudes and directions. *Assume* the sense of the vectors representing the *reactions acting on the assembly* (the correct sense will always emerge from the equilibrium equations for the assembly). Include any other relevant information e.g. lengths, angles etc which may help when formulating the equilibrium equations (including the moment equation) for the assembly.

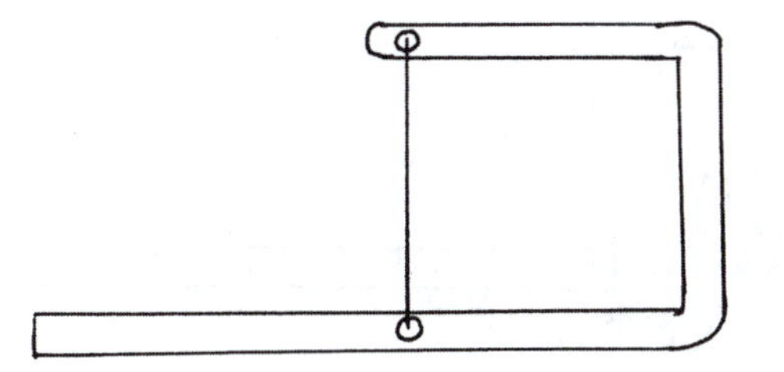

Problem 3.22.

In the following figure, the tension in cable AB is T. Draw a free-body diagram of the assembly and cable AB, treating them as a single object. Neglect the weight of the pipe assembly.

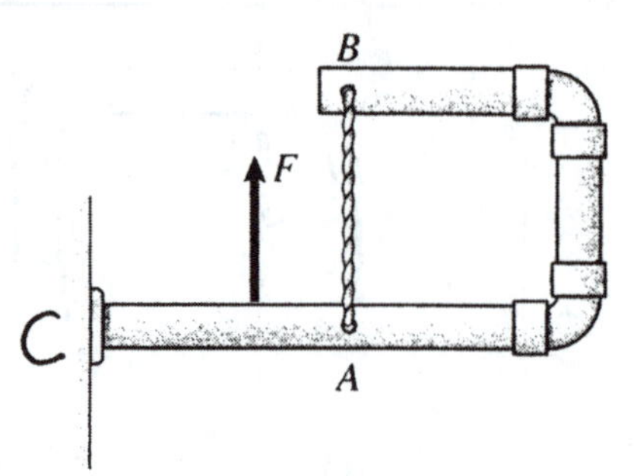

Solution

1. Imagine the entire pipe assembly to be separated or detached from the system (the supporting wall).
2. There is a built-in (fixed) support at C. Use Table 2.1 to determine the number and types of reactions *acting on the assembly* at C.
3. In addition to the force with magnitude F shown in the figure, the assembly is subjected to two additional *external* forces and an external couple. They are caused by:
   THE SUPPORT AT C
   There are no other external forces acting on the assembly.
4. Draw the free-body diagram of the (detached) assembly showing all these forces and any couples labeled with their magnitudes and directions. *Assume* the sense of the vectors representing the *reactions acting on the assembly* (the correct sense will always emerge from the equilibrium equations for the assembly). Include any other relevant information e.g. lengths, angles etc which may help when formulating the equilibrium equations (including the moment equation) for the assembly.

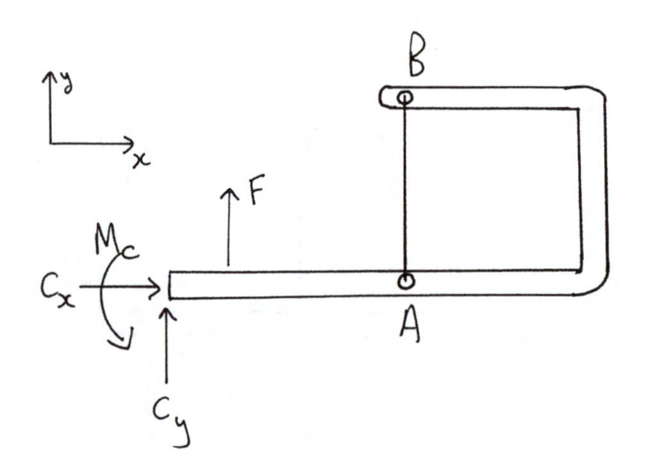

Problem 3.23.

Draw the free-body diagram of the assembly in **Problem 3.22** but this time *not* including cable AB.

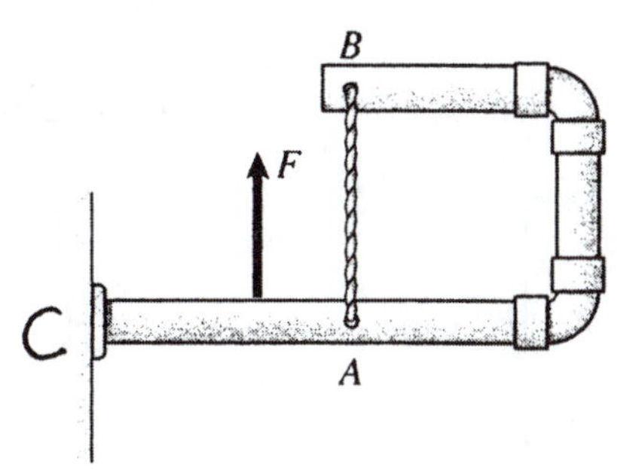

Solution

1. Imagine the assembly ABC to be separated or detached from the system (supporting wall + cable).
2. There is a built-in (fixed) support at C. As in Problem 3.22, use Table 2.1 to determine the number and types of reactions *acting on the assembly* at C. In addition to these reactions, we now have additional external forces acting on the assembly as a result of the tension in cable AB.
3. In addition to the force with magnitude F shown in the figure, the assembly is now, therefore, subjected to four *external* forces and an external couple. They are:

   i. **ii.**

   iii. **iv.**

   v.
4. Draw the free-body diagram of the (detached) assembly showing all these forces and any couples labeled with their magnitudes and directions. *Assume* the sense of the vectors representing the *reactions acting on the assembly* (the correct sense will always emerge from the equilibrium equations for the assembly). Include any other relevant information e.g. lengths, angles etc which may help when formulating the equilibrium equations (including the moment equation) for the assembly.

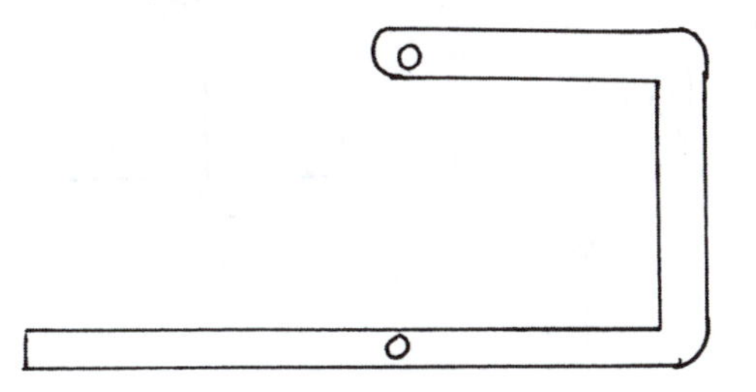

Problem 3.23.

Draw the free-body diagram of the assembly in **Problem 3.22** but this time *not* including cable AB.

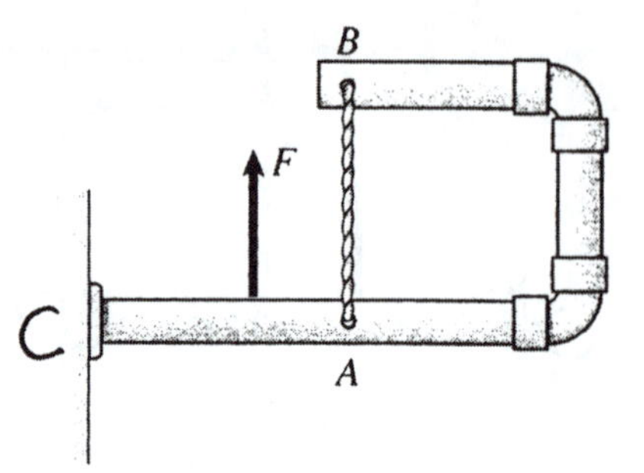

Solution

1. Imagine the assembly ABC to be separated or detached from the system (supporting wall + cable).
2. There is a built-in (fixed) support at C. As in Problem 3.22, use Table 2.1 to determine the number and types of reactions *acting on the assembly* at C. In addition to these reactions, we now have additional external forces acting on the assembly as a result of the tension in cable AB.
3. In addition to the force with magnitude F shown in the figure, the assembly is now, therefore, subjected to four *external* forces and an external couple. They are:

   i. (i) **REACTION AT** C **(2 forces and a couple)**

   ii. **TWO TENSION (CABLE) FORCES (equal magnitude, opposite directions) ACTING ON THE BAR AT** A and B.
4. Draw the free-body diagram of the (detached) assembly showing all these forces and any couples labeled with their magnitudes and directions. *Assume* the sense of the vectors representing the *reactions acting on the assembly* (the correct sense will always emerge from the equilibrium equations for the assembly). Include any other relevant information e.g. lengths, angles etc which may help when formulating the equilibrium equations (including the moment equation) for the assembly.

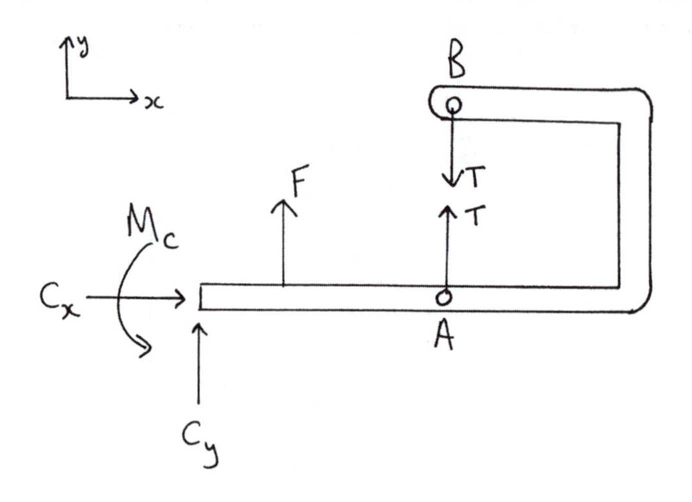

Problem 3.24.

Draw the free-body diagram of cable AB in the assembly of **Problem 3.22**.

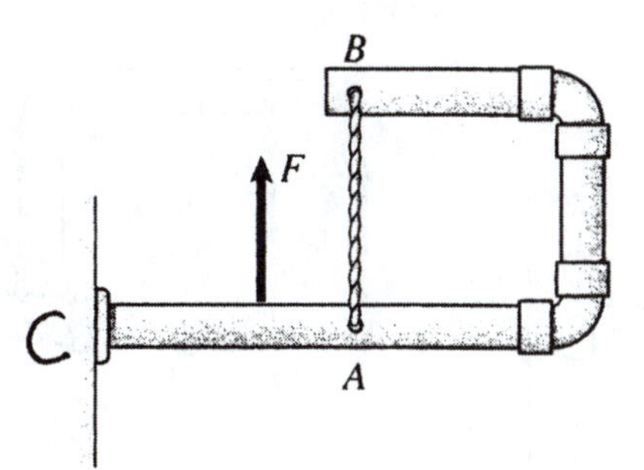

Solution

1. Imagine the cable to be separated or detached from the system.
2. The cable AB is subjected to two *external* forces. They are:

   i. **ii.**

3. Draw the free-body diagram of the (detached) cable showing these forces labeled with their magnitudes and directions.

Problem 3.24.

Draw the free-body diagram of cable AB in the assembly of **Problem 3.22**.

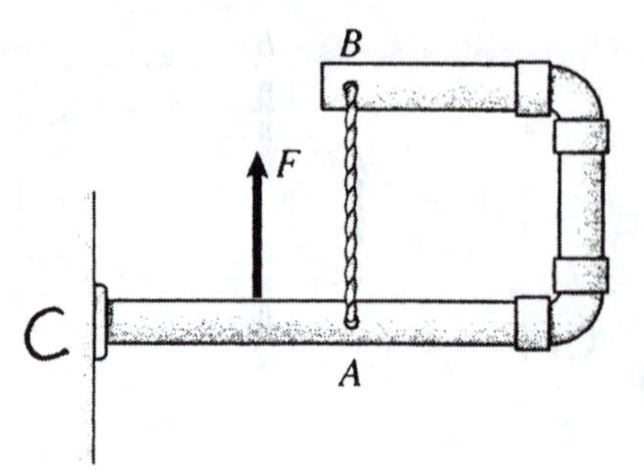

Solution

1. Imagine the cable to be separated or detached from the system.
2. The cable AB is subjected to two *external* forces. They are:
   i. **TWO TENSION FORCES (equal magnitude, opposite directions) ACTING ON THE CABLE AT** A and B. (These forces should be equal and opposite to the tension forces mentioned in Problem 3.23)
3. Draw the free-body diagram of the (detached) cable showing these forces labeled with their magnitudes and directions.

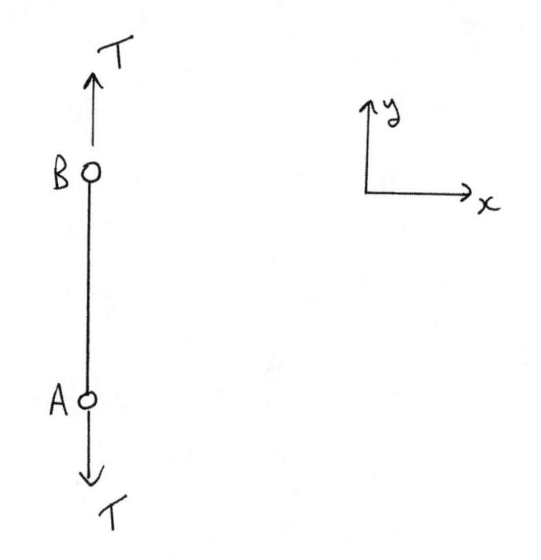

Problem 3.25.

The mass of the person is 80 kg and the mass of the diving board is 45 kg. Draw the free-body diagram of the diving board. Use the free-body diagram to write down equilibrium equations for the board and and to determine the reactions at the supports A and B.

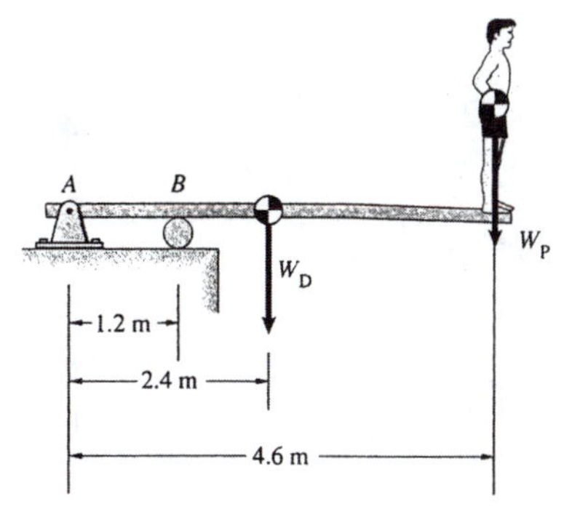

Solution

1. Imagine the diving board to be separated or detached from the system.
2. There is a pin support at A and a roller support at B. Use Table 2.1 to determine the number and types of reactions *acting on the board* at A and B.
3. The board is subjected to five *external* forces. They are caused by:

   i. **ii.**

   iii. **iv.**

   v.
4. Draw the free-body diagram of the (detached) board showing all these forces labeled with their magnitudes and directions. *Assume* the sense of the vectors representing the *reactions acting on the board* (the correct sense will always emerge from the equilibrium equations for the board). Include any other relevant information e.g. lengths, angles etc which may help when formulating the equilibrium equations (including the moment equation) for the board.
5. The weight of the person is: $W_P =$

   The weight of the diving board is: $W_D =$
6. Establish an xy-axes system on the free-body diagram and write down the force equilibrium equations in each of the x and $y-$ directions:

   $+\rightarrow \sum F_x = 0:$

   $+\uparrow \ \sum F_y = 0:$

   Sum moments about A and write down the moment equilibrium equation.

   $\curvearrowleft + \sum M_A = 0:$
7. Solve the three equations (using the values for W_P and W_D obtained above) for the required reactions at A and B.

Problem 3.25.

The mass of the person is 80 kg and the mass of the diving board is 45 kg. Draw the free-body diagram of the diving board. Use the free-body diagram to write down equilibrium equations for the board and and to determine the reactions at the supports A and B.

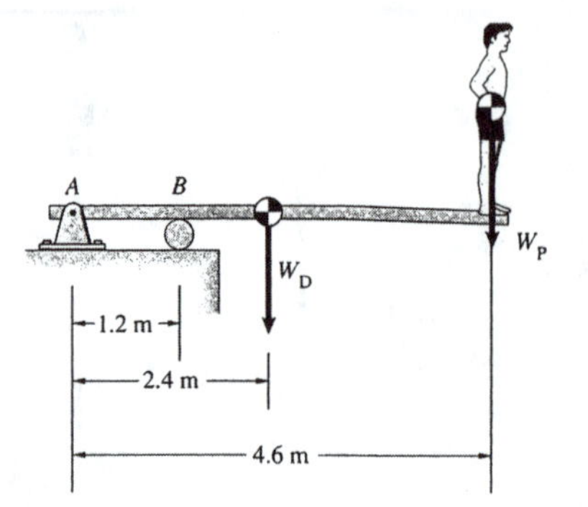

Solution

1. Imagine the diving board to be separated or detached from the system.
2. There is a pin support at A and a roller support at B. Use Table 2.1 to determine the number and types of reactions *acting on the board* at A and B.
3. The board is subjected to five *external* forces. They are caused by:

   i. REACTION AT A **(Two forces)** | **ii. REACTION AT** B

   iii. WEIGHT OF PERSON (W_P) | **iv. WEIGHT OF DIVING BOARD (W_D)**
4. Draw the free-body diagram of the (detached) board showing all these forces labeled with their magnitudes and directions. *Assume* the sense of the vectors representing the *reactions acting on the board* (the correct sense will always emerge from the equilibrium equations for the board). Include any other relevant information e.g. lengths, angles etc which may help when formulating the equilibrium equations (including the moment equation) for the board.

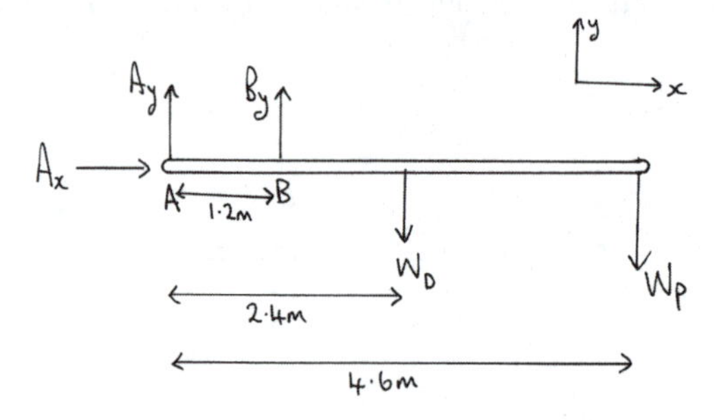

5. The weight of the person has magnitude: $W_P = mg = 80\,(9.8) = 784N$
   The weight of the diving board has magnitude: $W_D = mg = 45\,(9.8) = 441N$
6. Establish an xy-axes system on the free-body diagram and write down the force equilibrium equations in each of the x and $y-$ directions:
   $+\rightarrow \sum F_x = 0: \quad A_x = 0.$

   $+\uparrow \ \sum F_y = 0: \quad -W_P - W_D + B_y + A_y = 0.$
   Sum moments about A and write down the moment equilibrium equation.
   $\curvearrowleft + \sum M_A = 0: \quad -4.6(784) - 2.4\,(441) + 1.2B_y = 0.$
7. Solve the three equations (using the values for W_P and W_D obtained above) for the required reactions at A and B : $B_y = 3887.3N$, $A_y = -2662.3N$ (directions as shown on the FBD).

Problem 3.26.

Draw the free-body diagram of the beam and use it to determine the reactions at the support A. Neglect the weight of the beam.

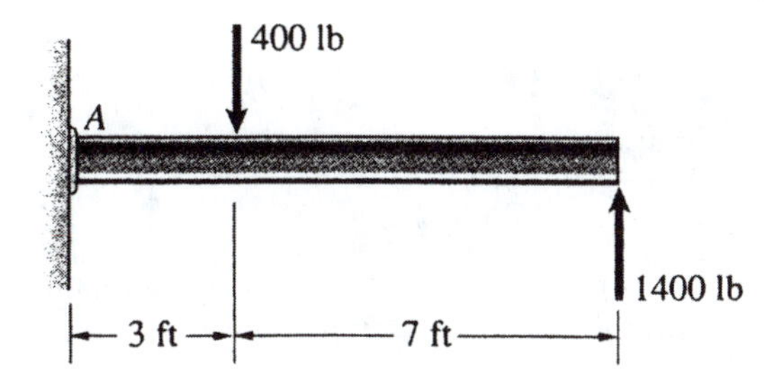

Solution

1. Imagine the beam to be separated or detached from the system (wall).
2. There is a built-in support at A. Use Table 2.1 to determine the number and types of reactions *acting on the beam* at A.
3. The beam is subjected to four *external* forces and an external couple. They are:

   i. **ii.**

   iii. **iv.**

   v.

4. Draw the free-body diagram of the (detached) beam showing all these forces and the external couple labeled with their magnitudes and directions. *Assume* the sense of the vectors representing the *reactions acting on the beam* (the correct sense will always emerge from the equilibrium equations for the beam). Include any other relevant information e.g. lengths, angles etc which may help when formulating the equilibrium equations (including the moment equation) for the beam.

5. Establish an xy-axes system on the free-body diagram and write down the force equilibrium equations in each of the x and $y-$ directions:

   $+\rightarrow \sum F_x = 0:$

   $+\uparrow \sum F_y = 0:$

   Sum moments about A and write down the moment equilibrium equation.

   $\curvearrowleft + \sum M_A = 0:$

6. Solve the three equations for the required reactions at the support A.

Problem 3.26.

Draw the free-body diagram of the beam and use it to determine the reactions at the support A. Neglect the weight of the beam.

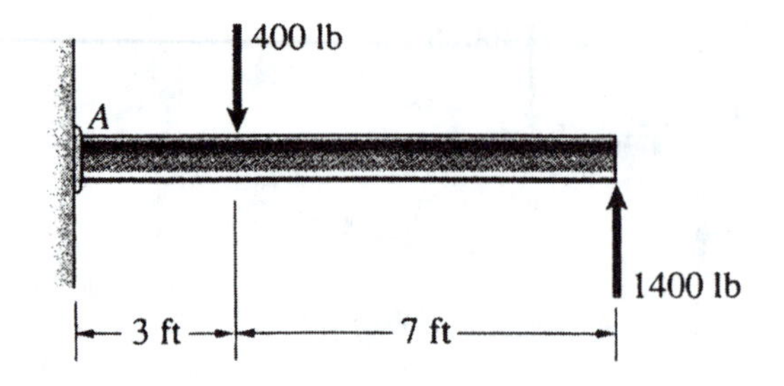

Solution

1. Imagine the beam to be separated or detached from the system (wall).
2. There is a built-in support at A. Use Table 2.1 to determine the number and types of reactions *acting on the beam* at A.
3. The beam is subjected to four *external* forces and an external couple. They are:

   i. REACTIONS AT A **(Two forces and a couple)** **ii.** 400 lb **LOAD**

   iii. 1400 lb **LOAD**

4. Draw the free-body diagram of the (detached) beam showing all these forces and the external couple labeled with their magnitudes and directions. *Assume* the sense of the vectors representing the *reactions acting on the beam* (the correct sense will always emerge from the equilibrium equations for the beam). Include any other relevant information e.g. lengths, angles etc which may help when formulating the equilibrium equations (including the moment equation) for the beam.

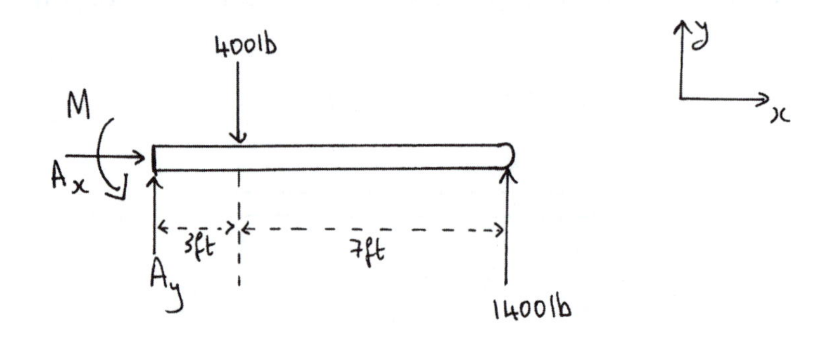

5. Establish an xy-axes system on the free-body diagram and write down the force equilibrium equations in each of the x and $y-$ directions:

   $+\rightarrow \sum F_x = 0: \quad A_x = 0.$

   $+\uparrow \ \sum F_y = 0: \quad A_y - 400 + 1400 = 0.$

   Sum moments about A and write down the moment equilibrium equation.

   $\curvearrowleft + \sum M_A = 0: \quad M - 3\,(400) + 10\,(1400) = 0$

6. Solve the three equations for the required reactions at the support A :

   $A_y = -1000$ lb, $M = -12800$ ft.lb (directions as shown on the FBD).

Problem 3.27.

Draw a free-body diagram of the beam.

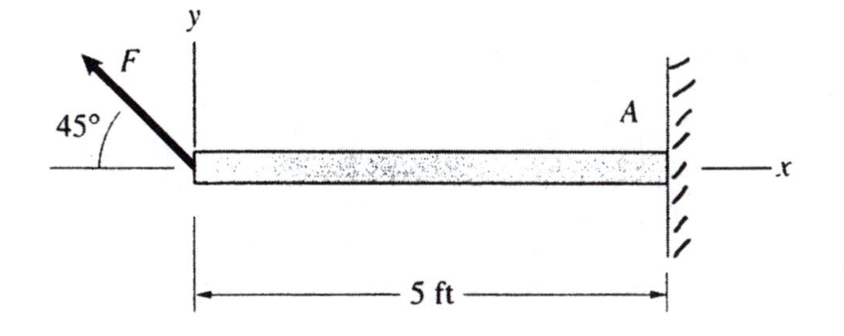

Solution

1. Imagine the beam to be separated or detached from the system.
2. There is a built-in (fixed) support at A. Use Table 2.1 to determine the number and types of reactions *acting on the beam* at A.
3. The beam is subjected to four *external* forces and one external couple. They are caused by:

4. Draw the free-body diagram of the (detached) beam showing all these forces and the couple labeled with their magnitudes and directions. *Assume* the sense of the vectors representing the *reactions acting on the beam.* Include any other relevant information e.g. lengths, angles etc which may help when formulating the equilibrium equations (including the moment equation) for the beam.

Problem 3.27.

Draw a free-body diagram of the beam.

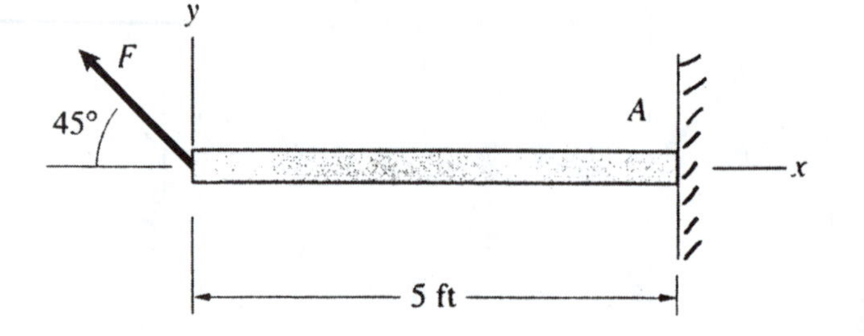

Solution

1. Imagine the beam to be separated or detached from the system.
2. There is a built-in (fixed) support at A. Use Table 2.1 to determine the number and types of reactions *acting on the beam* at A.
3. The beam is subjected to four *external* forces and one external couple. They are caused by:
   i. **REACTION AT** A **(Two forces and a couple).**
   ii. **APPLIED FORCE F.**
   iii. **WEIGHT OF BAR.**
4. Draw the free-body diagram of the (detached) beam showing all these forces and the couple labeled with their magnitudes and directions. *Assume* the sense of the vectors representing the *reactions acting on the beam.* Include any other relevant information e.g. lengths, angles etc which may help when formulating the equilibrium equations (including the moment equation) for the beam.

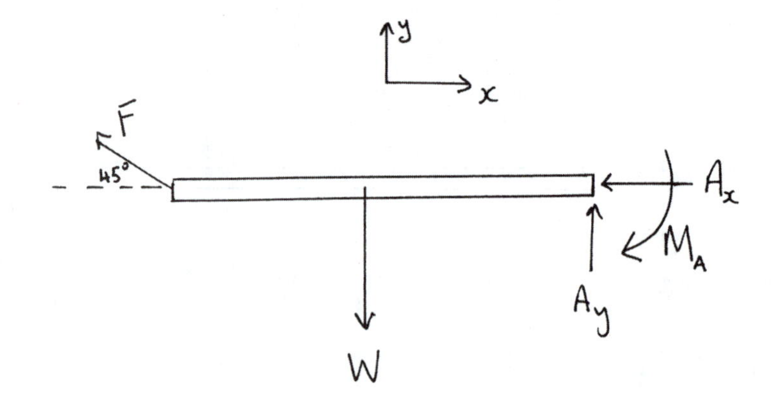

Problem 3.28.

The mobile is in equilibrium. The weights of the crossbars are negligible. Draw free-body diagrams for each of the crossbars.

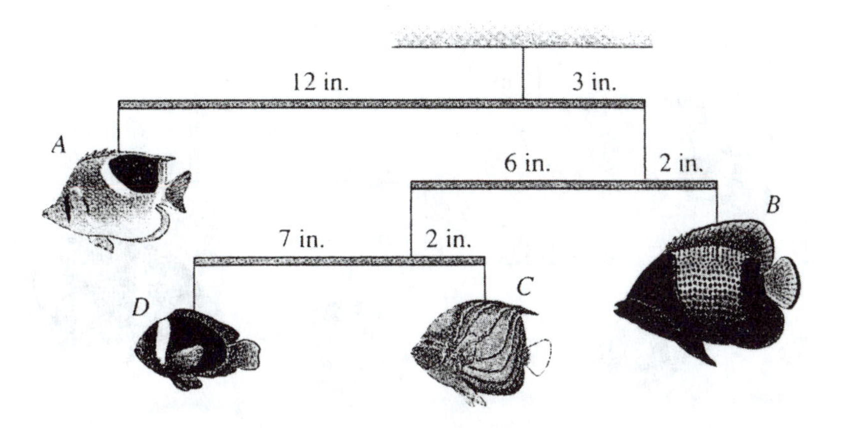

Solution

1. Imagine each crossbar to be separated or detached from the system.
2. Each crossbar has the reaction from a cable support and two weights acting externally. Be careful - not all of these forces are independent!
3. Draw the free-body diagram of each (detached) crossbar showing all these forces labeled with their magnitudes and directions. Include any other relevant information e.g. lengths, angles etc which may help when formulating the equilibrium equations (including the moment equation) for each crossbar.

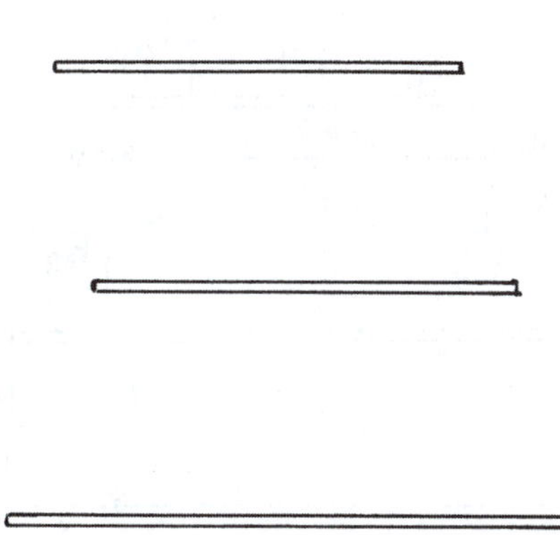

Problem 3.28.

The mobile is in equilibrium. The weights of the crossbars are negligible. Draw free-body diagrams for each of the crossbars.

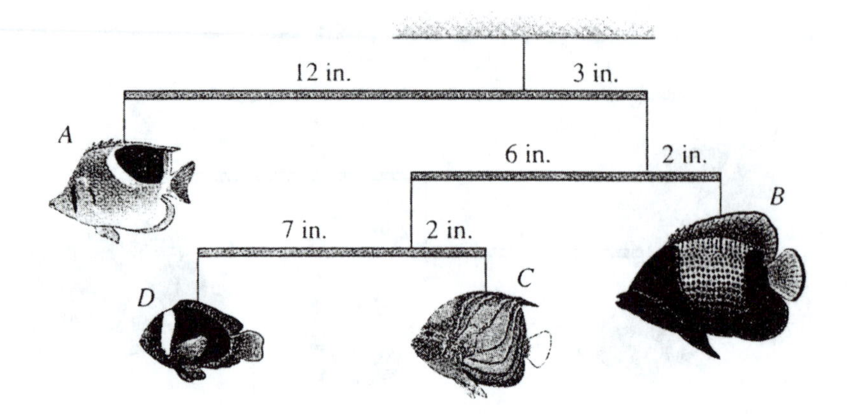

Solution

1. Imagine each crossbar to be separated or detached from the system.
2. Each crossbar has the reaction from a cable support and two weights acting externally. Be careful - not all of these forces are independent!
3. Draw the free-body diagram of each (detached) crossbar showing all these forces labeled with their magnitudes and directions. Include any other relevant information e.g. lengths, angles etc which may help when formulating the equilibrium equations (including the moment equation) for each crossbar.

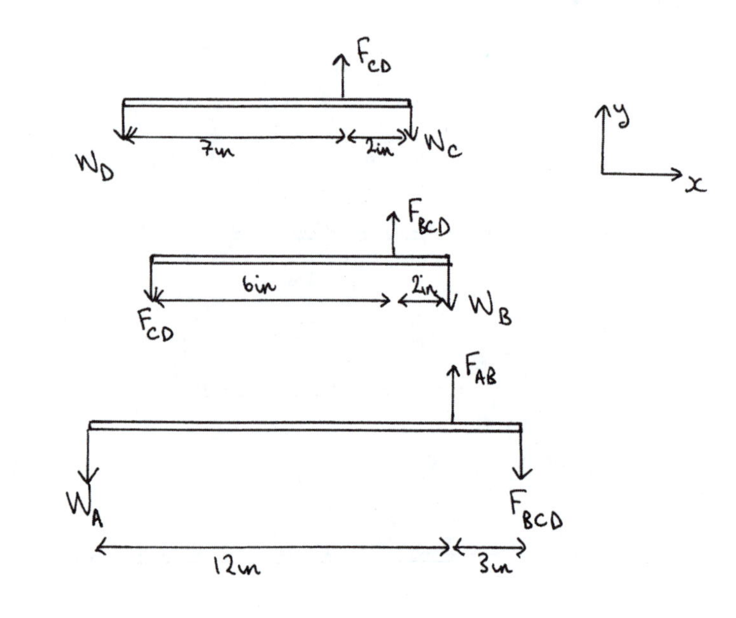

F_{AB}, F_{CD}, F_{BCD} : Magnitudes of reactions at the supports

W_A, W_B, W_C, W_D : Weights of A, B, C and D, respectively.

Problem 3.29.

The lift forces on an airplane's wing are represented by eight forces. The magnitude of each force is given in terms of its position x on the wing by $F_L = 200\sqrt{1-(x/17)^2}$ lb. The weight of the wing is $W = 400$ lb. Draw a free-body diagram for the wing and use it to formulate equilibrium equations for the wing. Solve these equilibrium equations to find the reactions on the wing at the root R.

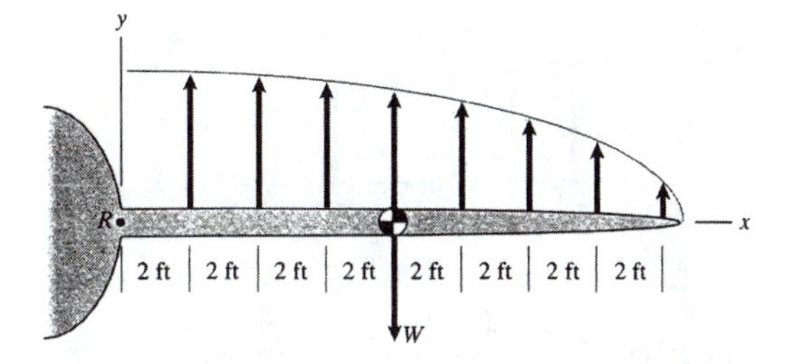

Solution

1. Imagine the wing to be separated or detached from the system (airplane).
2. There is a built-in support at R. Use Table 2.1 to determine the number and types of reactions *acting on the wing* at R.
3. Apart from the lift forces, the wing is subjected to three *external* forces and an external couple. They are caused by:
4. Draw the free-body diagram of the (detached) wing showing all these forces and the external couple labeled with their magnitudes and directions. *Assume* the sense of the vectors representing the *reactions acting on the wing* (the correct sense will always emerge from the equilibrium equations for the wing). Include any other relevant information e.g. lengths, angles etc which may help when formulating the equilibrium equations (including the moment equation) for the wing.

5. Make a table showing the magnitudes of the forces and the moments about the wing root R acting on the wing.

Station	Distance D (ft.)	Force Magnitude F_L	Moment Magnitude M_L
1	2		
2	4		
3	6		
4	8		
5	10		
6	12		
7	14		
8	16		
Sum			

6. Establish an $xy-$axes system on the free-body diagram and write down the force equilibrium equations in each of the x and $y-$ directions:

   $+\rightarrow \sum F_x = 0:$

   $+\uparrow \ \sum F_y = 0:$

   Sum moments about R and write down the moment equilibrium equation.

   $\curvearrowleft + \sum M_R = 0:$

7. Solve the three equations for the required reactions at the support R.

Problem 3.29.

The lift forces on an airplane's wing are represented by eight forces. The magnitude of each force is given in terms of its position x on the wing by $F_L = 200\sqrt{1-(x/17)^2}$ lb. The weight of the wing is $W = 400\ lb$. Draw a free-body diagram for the wing and use it to formulate equilibrium equations for the wing. Solve these equilibrium equations to find the reactions on the wing at the root R.

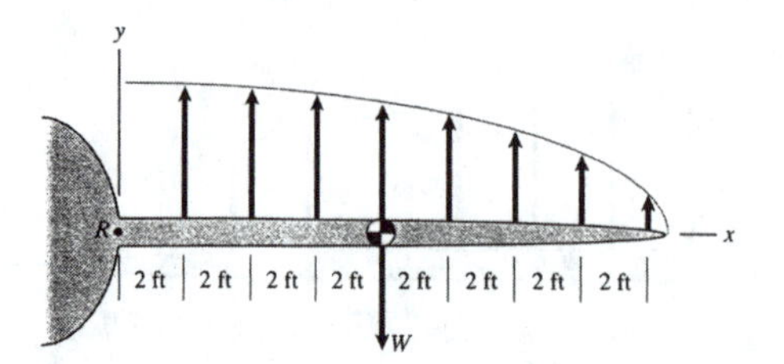

Solution

1. Imagine the wing to be separated or detached from the system (airplane).
2. There is a built-in support at R. Use Table 2.1 to determine the number and types of reactions *acting on the wing* at R.
3. Apart from the lift forces, the wing is subjected to three *external* forces and an external couple. They are caused by:
   i. **THE BUILT-IN SUPPORT AT R (two forces and a couple).**
   ii. **WEIGHT OF THE WING.**
4. Draw the free-body diagram of the (detached) wing showing all these forces and the external couple labeled with their magnitudes and directions. *Assume* the sense of the vectors representing the *reactions acting on the wing* (the correct sense will always emerge from the equilibrium equations for the wing). Include any other relevant information e.g. lengths, angles etc which may help when formulating the equilibrium equations (including the moment equation) for the wing.

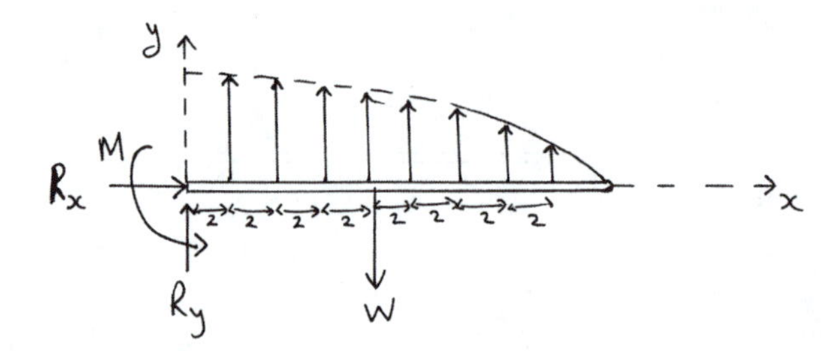

5. Make a table showing the magnitudes of the forces and the moments about the wing root R acting on the wing.

Station	Distance D (ft.)	Force Magnitude F_L	Moment Magnitude M_L
1	2	198.61	397.22
2	4	194.38	777.54
3	6	187.13	1122.78
4	8	176.47	1411.76
5	10	161.74	1617.38
6	12	141.67	1700.0
7	14	113.45	1588.37
8	16	67.58	1081.33
Sum $\sum$		1241.03	9696.40

6. Establish an $xy-$axes system on the free-body diagram and write down the force equilibrium equations in each of the x and $y-$ directions:
   $+\rightarrow \sum F_x = 0: \quad R_x = 0.$
   $+\uparrow \ \sum F_y = 0: \quad R_y + \sum F_L - W = 0.$
   Sum moments about R and write down the moment equilibrium equation.
   $\curvearrowleft + \sum M_R = 0: \quad M + \sum M_L - 8W = 0.$
7. Solve the three equations for the required reactions at the support R :
   $R_x = 0$, $R_y = -841$ lb, $M = -6496.3$ ft.lb (directions as shown on FBD).

Problem 3.30.

The weight W of the bar acts at its center. The surfaces are smooth. Draw a free-body diagram of the bar and use it to write down equilibrium equations for the bar.

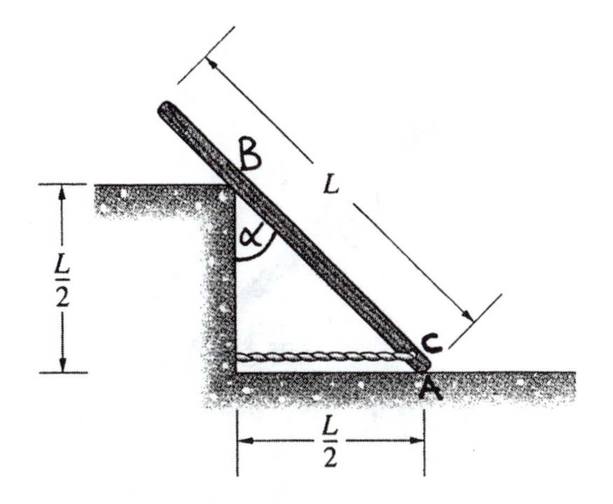

Solution

1. Imagine the bar to be separated or detached from the system.
2. The supports at A and B are contacts with a smooth surface. There is also a cable support at C. Use Table 2.1 to identify the reactions *acting on the bar* at A, B and C.
3. The bar is subjected to four *external* forces.
4. Draw the free-body diagram of the (detached) bar showing all these forces labeled with their magnitudes and directions. *Assume* the sense of the vectors representing the *reactions acting on the bar*. Include any other relevant information e.g. lengths, angles etc which may help when formulating the equilibrium equations (including the moment equation) for the bar.
5. On the free-body diagram, establish an $xy-$axes system and resolve all forces into x and y components.

6. Write down the force equilibrium equations in each of the x and $y-$ directions

$+\rightarrow \sum F_x = 0:$

$+\uparrow \ \sum F_y = 0:$

7. Sum moments about A and write down the moment equilibrium equation.

$\curvearrowleft + \sum M_A = 0:$

Problem 3.30.

The weight W of the bar acts at its center. The surfaces are smooth. Draw a free-body diagram of the bar and use it to write down equilibrium equations for the bar.

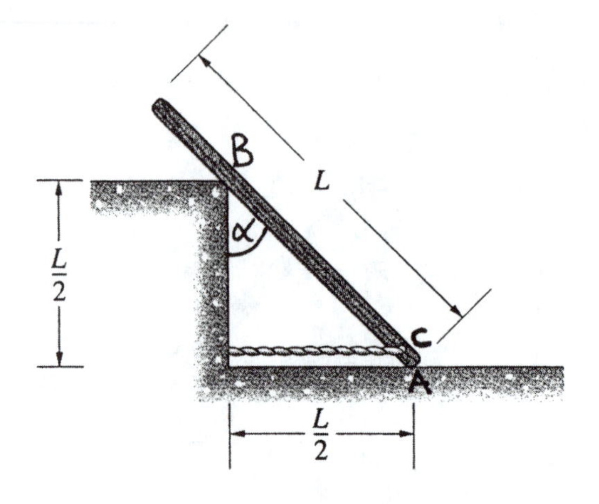

Solution

1. Imagine the bar to be separated or detached from the system.
2. The supports at A and B are contacts with a smooth surface. There is also a cable support at C. Use Table 2.1 to identify the reactions *acting on the bar* at A, B and C.
3. The bar is subjected to four *external* forces.
4. Draw the free-body diagram of the (detached) bar showing all these forces labeled with their magnitudes and directions. *Assume* the sense of the vectors representing the *reactions acting on the bar*. Include any other relevant information e.g. lengths, angles etc which may help when formulating the equilibrium equations (including the moment equation) for the bar.
5. On the free-body diagram, establish an $xy-$axes system and resolve all forces into x and y components.

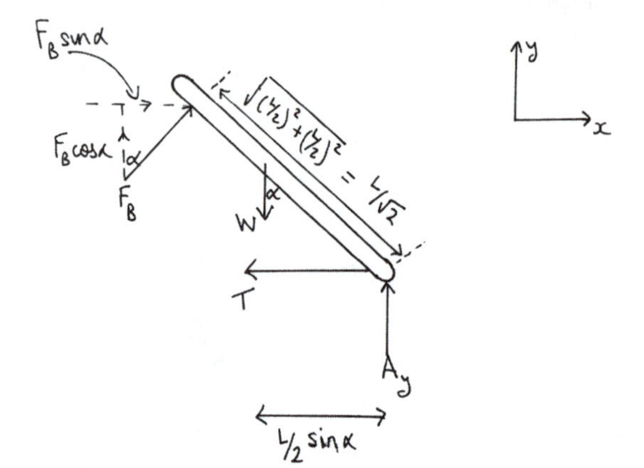

6. Write down the force equilibrium equations in each of the x and $y-$ directions (noting that $\alpha = 45°$):

$$+\rightarrow \sum F_x = 0: \quad -T + F_B \sin\alpha = 0 \Leftrightarrow -T + \frac{F_B}{\sqrt{2}} = 0.$$

$$+\uparrow \sum F_y = 0: \quad -W + A_y + F_B \cos\alpha = 0 \Leftrightarrow -W + A_y + \frac{F_B}{\sqrt{2}} = 0.$$

7. Sum moments about A and write down the moment equilibrium equation.

$$\curvearrowleft + \sum M_A = 0: \quad \frac{WL}{2}\sin\alpha - \frac{F_B L}{\sqrt{2}} = 0 \Leftrightarrow \frac{WL}{2\sqrt{2}} + \frac{F_B L}{\sqrt{2}} = 0.$$

PART II

SECTION-BY-SECTION, CHAPTER-BY-CHAPTER SUMMARIES WITH REVIEW QUESTIONS AND ANSWERS

1

Introduction

Main Goals of this Chapter:

- To introduce the basic ideas of *Mechanics.*
- To outline a general procedure for solving problems in mechanics.
- To review fundamental concepts in mechanics.
- To review the principles for applying the SI and U.S. Customary systems of units.
- To give a concise statement of Newton's theory of gravitation.

1.1 Engineering and Mechanics

Modern engineers use *mathematical models* to predict and analyze the behavior of complex engineering systems. As an analytical science, *mechanics* is one of the main sources of fundamental concepts and analytical methods used in the design and analysis of mathematical models.

Mechanics is that branch of the physical sciences concerned with the behavior of bodies subjected to the action of forces. The subject of mechanics is divided into two parts:

- *statics*—the study of objects in equilibrium (objects either at rest or moving with a constant velocity).
- *dynamics*—the study of objects with accelerated motion.

Although statics can be considered as a special case of dynamics (in which the acceleration is zero), it deserves special treatment since many objects are designed with the intention that they remain in equilibrium.

Problem Solving

- The most effective way to learn engineering mechanics is to *solve problems.* The following problem-solving procedure will be useful:
- Problem Solving
  - — Identify the information that is given and the information, or answer, you must determine. It's often helpful to restate the problem in your own words. When appropriate, make sure you understand the physical system or model involved.

— Develop a strategy for the problem. This means identifying the principles and equations that apply and deciding how you will use them to solve the problem. Whenever possible, draw diagrams to help visualize and solve the problem.

— Whenever you can, try to predict the answer. This will develop your intuition and will often help you recognize an incorrect answer.

— Solve the equations, and, whenever possible, interpret your results and compare them with your prediction. This last step is a *reality check*. Is your answer reasonable?

Fundamental Concepts

Numbers

- *Significant Digits*. This term refers to the number of meaningful (accurate) digits in a number, counting to the right starting with the first nonzero digit. The two numbers 7.630 and 0.007630 are each stated to four significant digits. If only the first four digits in the number 7,630,000 are known to be accurate, this can be indicated by writing the number in scientific notation as 7.630×10^6.

Space and Time

- *Space* simply refers to the three-dimensional universe in which we live. The distance between two points in space is the *length* of the straight line joining them. In SI units, the unit of length is the meter (m). In U.S. Customary units, the unit of length is the foot (ft).
- *Time* is measured by the intervals between repeatable events, such as the swings of a clock pendulum. In both SI and U.S. Customary units, the unit of time is the second (s).
- If the position of a point is space relative to some reference point changes with time, the rate of change of its position is called its *velocity*, and the rate of change of its velocity is called its *acceleration*. In SI units, the velocity is measured in *meters per second* (m/s) and the acceleration in *meters per second per second, or meters per second squared* (m/s^2). In *U.S. Customary* units, the velocity is expressed in *feet per second* (ft/s) and the acceleration is expressed in *feet per second squared* (ft/s^2).

Newton's Laws

Newton's laws apply to the motion of a particle as measured from a nonaccelerating (inertial) reference frame.

* **First Law.** *When the sum of the forces acting on a particle is zero, its velocity is constant. In particular, if the particle is initially stationary, it will remain stationary.*
* **Second Law.** *When the sum of the forces acting on a particle is non-zero, the sum of the forces is equal to the rate of change of the linear momentum of the particle. If the mass is constant, the sum of the forces is equal to the product of the mass of the particle and its acceleration.* i.e.,

$$\mathbf{F} = m\mathbf{a}$$

* **Third Law**. *The forces exerted by two particles on each other are equal in magnitude and opposite in direction.*

Note: Newton's second law gives precise meanings to the terms *mass* and *force*. In SI units, the unit of mass is the *kilogram* (kg). The unit of force is the *Newton* (N), which is the force required to give a mass of one kilogram an acceleration of one meter per second squared. In *U.S. Customary* units, the unit of force is the *pound* (lb). The unit of mass is the *slug*, which is the amount of mass accelerated at one foot per second squared by a force of one pound.

Units

The four basic quantities *force, mass, length* and *time* are related by Newton's second law. Hence, the units used to define these quantities are not independent i.e., three of the four units are called *base units* (arbitrarily defined) and the fourth unit is called a *derived unit* (derived from Newton's second law).

International System of Units (SI Units)

- In the SI system, the unit of force, the *newton*, is a derived unit. The meter, second, and kilogram are base units.
- One *newton* is equal to a force required to give one kilogram of mass an acceleration of 1 m/s^2.
- The common prefixes used in SI units are:

Prefix	Abbreviation	Multiple
nano-	n	10^{-9}
micro-	μ	10^{-6}
milli-	m	10^{-3}
kilo-	k	10^3
mega-	M	10^6
giga-	G	10^9

U.S. Customary

- In the U.S. Customary system, the unit of mass, the *slug*, is a derived unit. The foot, second and pound are base units.
- One *slug* is equal to the amount of matter accelerated at 1 ft/s^2 when acted upon by a force of 1 lb.

The following table summarizes the two systems of units.

Name	**Length**	**Time**	**Mass**	**Force**
International System (SI)	*meter(m)*	*second(s)*	*kilogram(kg)*	*newton** $\left(N = \frac{kg \cdot m}{s^2}\right)$
U.S. Customary	*foot(ft)*	*second(s)*	*slug** $\left(= \frac{lb \cdot s^2}{ft}\right)$	*pound(lb)*
**Derived Unit*				

Angular Units

- In both SI and $U.S.$ Customary units, angles are normally expressed in *radians(rad)*. An angle of 360° equals 2π radians.

- Equations containing angles are nearly always derived under the assumption that angles are expressed in radians. Therefore, when you want to substitute the value of an angle expressed in degrees into an equation, you should first convert it into radians.

Conversion of Units

The following table provides a set of direct conversion factors between U.S. Customary and SI units for the basic quantities.

Quantity	Unit (U.S. Customary)	Equals	Unit (SI)
Force	*lb*		4.448 N
Mass	*slug*		14.59 kg
Length	*ft*		0.3048 m
Length	*in*		25.4 mm

Note also that in the U.S. Customary system

$1\ ft = 12\ in. (inches).$
$5280\ ft = 1\ mi\ (mile).$
$1000\ lb = 1\ kip\ (1 kilo\text{-}pound).$
$2000\ lb = 1\ ton$

1.2 Newtonian Gravitation

$$F = G\frac{m_1 m_2}{r^2}$$

$$F = \text{force of gravitation between two particles}$$

$$G = \text{universal constant of gravitation}$$

$$m_1, m_2 = \text{mass of each of the two particles}$$

$$r = \text{distance between the two particles}$$

Note: The weight of a body has magnitude

$$W = mg,$$

where g is the acceleration due to gravity at sea level. The value of g varies from location to location on the surface of the earth but here is taken to be $g = 9.81\ m/s^2$ in SI units and $g = 32.2\ ft/s^2$ in U.S. Customary units.

Helpful Tips and Suggestions

- The *language* of engineering mechanics is *mathematics*. Consequently, make sure you review/reread the necessary mathematical notation/concepts *as they arise* in your mechanics course (trying to review all of the necessary mathematics *at once* is not recommended—there's just too much to digest at one time). You should aim to achieve *fluency* in basic mathematical techniques/notation so that your learning of mechanics is not distracted by trying to remember things which your instructor *assumes* you know, e.g., how to solve linear systems of algebraic equations, how to perform basic vector algebra, differentiation, and integration etc.

- *Remember* that in solving problems from engineering mechanics you are solving real, practical problems and producing real data with physical significance. Thus, you are responsible for making sure your results are correct, consistent, and well-presented. Get into the habit of doing this *now* so that it will become second nature by the time you graduate. In the world of professional engineering, you have a responsibility to your profession and to the many people that will use the product you will help to design, manufacture, or implement.

Review Questions: True or False

1. The subject called *Statics* studies only bodies which are at rest.
2. The best way to learn mechanics is to solve relevant problems.
3. The numbers 3.456 and 0.003456 are each stated to 4 significant digits.
4. Newton's second law can be written as $\mathbf{F} = m\mathbf{a}$ irrespective of whether mass is constant or varies with time.
5. In the *SI* system of units, the *newton* is a derived unit.
6. In the *U.S.* Customary system of units, the *pound* (*lb*) is the derived unit.
7. It's OK to mix degrees with radians in any equation involving angles.
8. Newtonian gravitation can be used to approximate the weight of an object of mass m due to the gravitational attraction of the earth.
9. Weight is a property of matter that does not change from one location to another.
10. If you know an object's mass, its weight at sea level can be determined by $W = mg$.

[0]1. F 2. T 3. T 4. F 5. T 6. F 7. F 8. T 9. F 10. T

2

Vectors

In this chapter we define scalars, vectors and fundamental vector operations. Specifically:

Main Goals of this Chapter:

- To define what we mean by *scalars* and *vectors*.
- To review rules for manipulating vectors.
- To express a vector and its properties in terms of its *scalar components* in two or three dimensions.
- To explain how to determine a vector's magnitude and direction.
- To introduce various *vector products* and their meanings.

2.1 Scalars and Vectors

Most of the physical quantities in mechanics can be represented by either *scalars* or *vectors*:

- A *scalar* a is a real number e.g., mass, time, volume, and length are represented by scalars.
- A *vector* $\mathbf{U}$ has both magnitude (a nonnegative real number denoted by $|\mathbf{U}|$) and direction e.g., force, velocity, and acceleration are vectors.

Rules for Manipulating Vectors

Vector Addition

- Consider two vectors $\mathbf{U}$ and $\mathbf{V}$. If we place them head to tail, their sum is defined to be the vector from the tail of $\mathbf{U}$ to the head of $\mathbf{V}$. This is called the *triangle rule* for vector addition. The sum is independent of the order in which the vectors are placed head to tail (*parallelogram rule*).

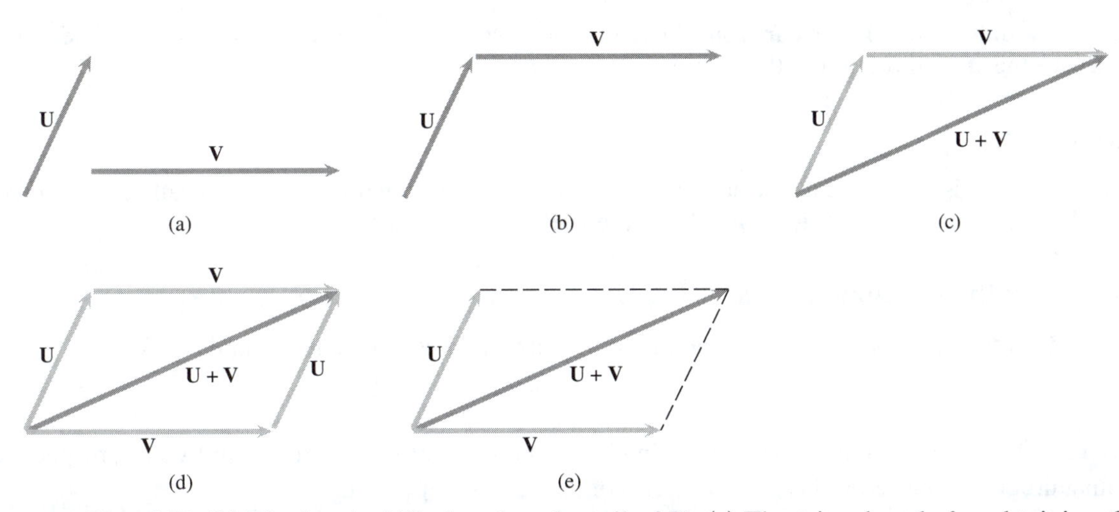

(a) Two vectors **U** and **V**. (b) The head of **U** placed at the tail of **V**. (c) The triangle rule for obtaining the sum of **U** and **V**. (d) The sum is independent of the order in which the vectors are added. (e) The parallelogram rule for obtaining the sum of **U** and **V**.

For any vectors **U**, **V**, and **W**, we have

$$\mathbf{U} + \mathbf{V} = \mathbf{V} + \mathbf{U} \quad \text{(commutative law)} \tag{2.1}$$

$$(\mathbf{U} + \mathbf{V}) + \mathbf{W} = \mathbf{U} + (\mathbf{V} + \mathbf{W}) \quad \text{(associative law).} \tag{2.2}$$

Product of a Scalar and a Vector

- The product of a vector **U** and a scalar a is a vector $a\mathbf{U}$ with magnitude $|a\mathbf{U}| = |a||\mathbf{U}|$. The direction is the same as that of **U** if a is positive and opposite to that of **U** if a is negative.
- For any scalars a and b and vectors **U** and **V**, we have

$$a(b\mathbf{U}) = (ab)\,\mathbf{U} \tag{2.3}$$

$$(a + b)\,\mathbf{U} = a\mathbf{U} + b\mathbf{U} \tag{2.4}$$

$$a\,(\mathbf{U} + \mathbf{V}) = a\mathbf{U} + a\mathbf{V}. \tag{2.5}$$

Vector Subtraction

- The difference of two vectors **U** and **V** is obtained by adding **U** to the vector $(-1)\mathbf{V}$:

$$\mathbf{U} - \mathbf{V} = \mathbf{U} + (-1)\,\mathbf{V}. \tag{2.6}$$

Unit Vectors

- A *unit vector* is simply a vector whose magnitude is 1. A unit vector specifies a direction and also provides a convenient way to express a vector that has a particular direction. For example, we can write

$$\mathbf{U} = |\mathbf{U}|\mathbf{e},$$

where **e** is a unit vector. It is clear from this equation that dividing any vector by its magnitude yields a unit vector with the same direction as the vector.

Vector Components

- When a vector **U** is expressed as the sum of a set of vectors, each vector of the set is called a *vector component* of **U**. We say that the vector **U** is *resolved* into the vector components.

2.2 Components in Two Dimensions

- A vector **U** in two dimensions is expressed in terms of its scalar components U_x and U_y as

$$\mathbf{U} = U_x\mathbf{i} + U_y\mathbf{j}, \tag{2.7}$$

where **i** and **j** are the usual unit vectors pointing in the direction of the positive x and y *axes*, respectively.

- The magnitude of the vector **U** is given in terms of its scalar components as

$$|\mathbf{U}| = \sqrt{U_x^2 + U_y^2}. \tag{2.8}$$

Manipulating Vectors in Terms of Their Components

- If $\mathbf{U} = (U_x\mathbf{i} + U_y\mathbf{j})$ and $\mathbf{V} = (V_x\mathbf{i} + V_y\mathbf{j})$,

$$\begin{aligned} \mathbf{U} + \mathbf{V} &= (U_x\mathbf{i} + U_y\mathbf{j}) + (V_x\mathbf{i} + V_y\mathbf{j}) \\ &= (U_x + V_x)\mathbf{i} + (U_y + V_y)\mathbf{j}. \end{aligned} \tag{2.9}$$

Also,

$$a\mathbf{U} = a(U_x\mathbf{i} + U_y\mathbf{j}) = aU_x\mathbf{i} + aU_y\mathbf{j}.$$

Position Vectors in Terms of Components

- We can express the position vector of a point relative to another point in terms of the Cartesian coordinates of the points. Consider points $A(x_A, y_A)$ and $B(x_B, y_B)$. The vector $\mathbf{r}_{AB}$ which specifies the position of B relative to A is given by

$$\mathbf{r}_{AB} = (x_B - x_A)\mathbf{i} + (y_B - y_A)\mathbf{j}. \tag{2.10}$$

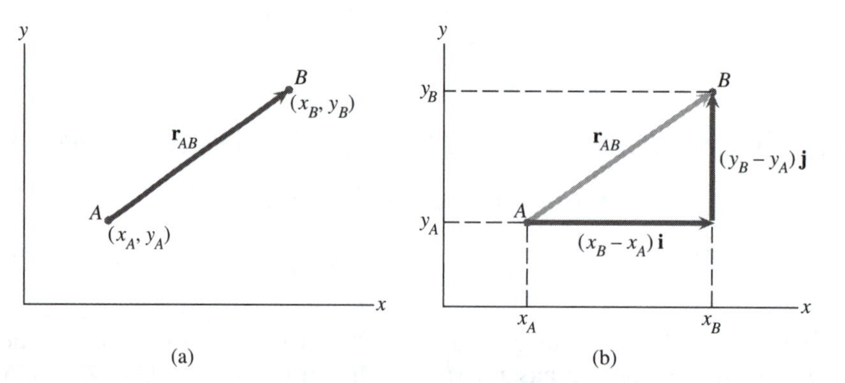

(a) Two points A and B and the position vector $\mathbf{r}_{AB}$ from A to B. (b) The components of $\mathbf{r}_{AB}$ can be determined from the coordinates of points A and B.

2.3 Components in Three Dimensions

- A vector **U** in three dimensions is expressed in terms of its scalar components U_x, U_y, and U_z as

$$\mathbf{U} = U_x\mathbf{i} + U_y\mathbf{j} + U_z\mathbf{k}, \tag{2.12}$$

where **i**, **j** and **k** are the usual unit vectors pointing in the direction of the positive x, y and z-*axes*, respectively, of a *right-handed coordinate system*: If the fingers of the right-hand are pointed in the positive x-*direction* and then closed toward the positive y-*direction*, the thumb points in the z-*direction*.

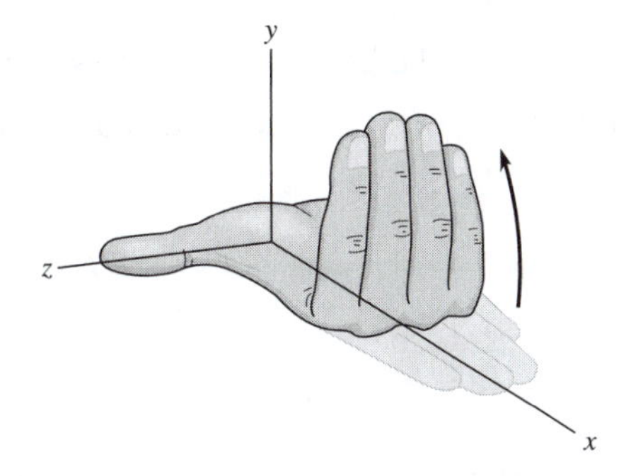

Recognizing a right-handed coordinate system.

- The magnitude of the vector **U** is given in terms of its scalar components as

$$|\mathbf{U}| = \sqrt{U_x^2 + U_y^2 + U_z^2}. \tag{2.14}$$

Direction Cosines

- Let θ_x, θ_y, and θ_z be the angles between the vector **U** and the positive coordinate axes

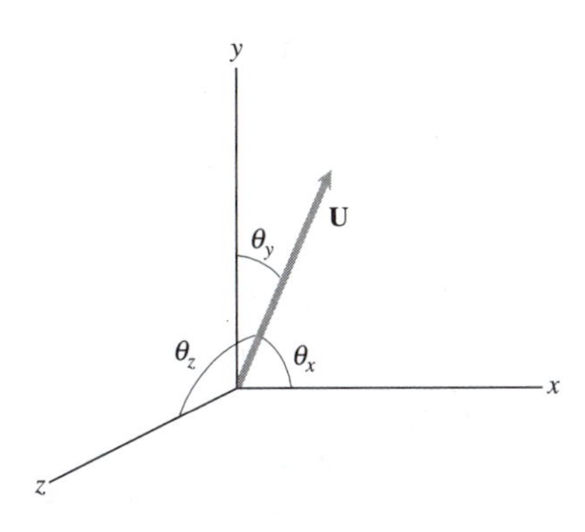

A vector U and the angles θ_x, θ_y, and θ_z

Then the scalar components of **U** can be written as

$$U_x = |\mathbf{U}|\cos\theta_x, \quad U_y = |\mathbf{U}|\cos\theta_y, \quad U_z = |\mathbf{U}|\cos\theta_z. \tag{2.15}$$

The quantities $\cos\theta_x$, $\cos\theta_y$ and $\cos\theta_z$ are the *direction cosines* of **U** and satisfy the relation

$$\cos^2\theta_x + \cos^2\theta_y + \cos^2\theta_z = 1. \tag{2.16}$$

Position Vectors in Terms of Components

- We can express the position vector of a point relative to another point in terms of the Cartesian coordinates of the points. Consider points $A(x_A, y_A, z_A)$ and $B(x_B, y_B, z_B)$. The vector $\mathbf{r}_{AB}$ which specifies the position of B relative to A is given by

$$\mathbf{r}_{AB} = (x_B - x_A)\,\mathbf{i} + (y_B - y_A)\mathbf{j} + (z_B - z_A)\,\mathbf{k}. \tag{2.17}$$

Components of a Vector Parallel to a Given Line

- If we know the magnitude of a vector **U** and the components of any vector **V** that has the same direction as **U**, then $\dfrac{\mathbf{V}}{|\mathbf{V}|}$ is a unit vector with the same direction as **U**, and we can determine the components of **U** by expressing it as $\mathbf{U} = |\mathbf{U}|\left(\dfrac{\mathbf{V}}{|\mathbf{V}|}\right)$.

2.4 Dot Products

Definition

- The dot (or scalar) product of two vectors **U** and **V** is

$$\mathbf{U}\cdot\mathbf{V} = |\mathbf{U}|\,|\mathbf{V}|\cos\theta, \tag{2.18}$$

where θ is the angle between the vectors **U** and **V** when they are placed tail to tail.
- *The dot product of two nonzero vectors is equal to zero if and only if the vectors are perpendicular.*
- The dot product has the following properties:

$$\mathbf{U}\cdot\mathbf{V} = \mathbf{V}\cdot\mathbf{U} \tag{2.19}$$

$$a\,(\mathbf{U}\cdot\mathbf{V}) = (a\mathbf{U})\cdot\mathbf{V} = \mathbf{U}\cdot(a\mathbf{V}) \tag{2.20}$$

$$\mathbf{U}\cdot(\mathbf{V}+\mathbf{W}) = \mathbf{U}\cdot\mathbf{V} + \mathbf{U}\cdot\mathbf{W}, \tag{2.21}$$

for any scalar a and any vectors **U**, **V**, and **W**.

Dot Product in Terms of Components

- It is also useful to note that

$$\begin{array}{lll} \mathbf{i}\cdot\mathbf{i}=1 & \mathbf{i}\cdot\mathbf{j}=0 & \mathbf{i}\cdot\mathbf{k}=0 \\ \mathbf{j}\cdot\mathbf{i}=0 & \mathbf{j}\cdot\mathbf{j}=1 & \mathbf{j}\cdot\mathbf{k}=0 \\ \mathbf{k}\cdot\mathbf{i}=0 & \mathbf{k}\cdot\mathbf{j}=0 & \mathbf{k}\cdot\mathbf{k}=1. \end{array} \tag{2.22}$$

- In terms of scalar components

$$\mathbf{U}\cdot\mathbf{V} = U_x V_x + U_y V_y + U_z V_z. \tag{2.23}$$

- To obtain the angle θ in terms of the components of the vectors:

$$\cos\theta = \frac{\mathbf{U}\cdot\mathbf{V}}{|\mathbf{U}|\,|\mathbf{V}|} = \frac{U_x V_x + U_y V_y + U_z V_z}{|\mathbf{U}|\,|\mathbf{V}|}. \tag{2.24}$$

Vector Components Parallel and Normal to a Line

- A vector $\mathbf{U}$ can be resolved into vector components $\mathbf{U}_P$ and $\mathbf{U}_n$ parallel and normal, respectively, to a straight line L. In terms of a unit vector $\mathbf{e}$ that is parallel to L, the parallel component, or projection of $\mathbf{U}$ onto L, is

$$\mathbf{U}_P = (\mathbf{e} \cdot \mathbf{U})\,\mathbf{e}. \tag{2.26}$$

Once the parallel component has been determined, the normal component $\mathbf{U}_n$ can be determined from the equation $\mathbf{U} = \mathbf{U}_n + \mathbf{U}_P$ i.e.

$$\mathbf{U}_n = \mathbf{U} - \mathbf{U}_P. \tag{2.27}$$

2.5 Cross Products

Definition

- The cross (or vector) product of two vectors $\mathbf{U}$ and $\mathbf{V}$ is

$$\mathbf{U} \times \mathbf{V} = |\mathbf{U}|\,|\mathbf{V}| \sin\theta\,\mathbf{e} \tag{2.28}$$

where θ is the angle between the vectors $\mathbf{U}$ and $\mathbf{V}$ when they are placed tail to tail and $\mathbf{e}$ is a unit vector perpendicular to $\mathbf{U}$ and $\mathbf{V}$. The direction of $\mathbf{e}$ is specified by the *right-hand rule:* When the fingers of the right hand are pointed in the direction of $\mathbf{U}$ (the first vector in the cross-product) and closed toward $\mathbf{V}$ (the second vector in the cross-product), the thumb points in the direction of $\mathbf{e}$.

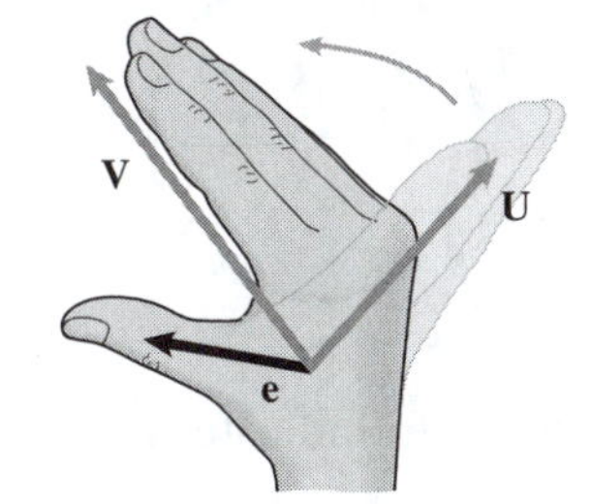

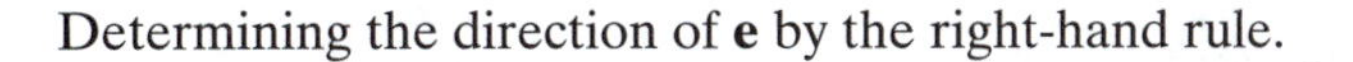
Determining the direction of $\mathbf{e}$ by the right-hand rule.

- *The cross-product of two nonzero vectors is equal to zero if and only if the two vectors are parallel.*
- The cross product has the following properties:

$$\mathbf{U} \times \mathbf{V} = -\mathbf{V} \times \mathbf{U} \tag{2.29}$$

$$a\,(\mathbf{U} \times \mathbf{V}) = (a\mathbf{U}) \times \mathbf{V} = \mathbf{U} \times (a\mathbf{V}) \tag{2.30}$$

$$\mathbf{U} \times (\mathbf{V} + \mathbf{W}) = (\mathbf{U} \times \mathbf{V}) + (\mathbf{U} \times \mathbf{W}), \tag{2.31}$$

for any scalar a and any vectors $\mathbf{U}$, $\mathbf{V}$ and $\mathbf{W}$.

Cross Product in Terms of Components

- The following useful results can be obtained by applying the right-hand rule and don't need to be memorized:

$$\begin{array}{lll} \mathbf{i} \times \mathbf{i} = \mathbf{0} & \mathbf{i} \times \mathbf{j} = \mathbf{k} & \mathbf{i} \times \mathbf{k} = -\mathbf{j} \\ \mathbf{j} \times \mathbf{i} = -\mathbf{k} & \mathbf{j} \times \mathbf{j} = \mathbf{0} & \mathbf{j} \times \mathbf{k} = \mathbf{i} \\ \mathbf{k} \times \mathbf{i} = \mathbf{j} & \mathbf{k} \times \mathbf{j} = -\mathbf{i} & \mathbf{k} \times \mathbf{k} = \mathbf{0}. \end{array} \quad (2.32)$$

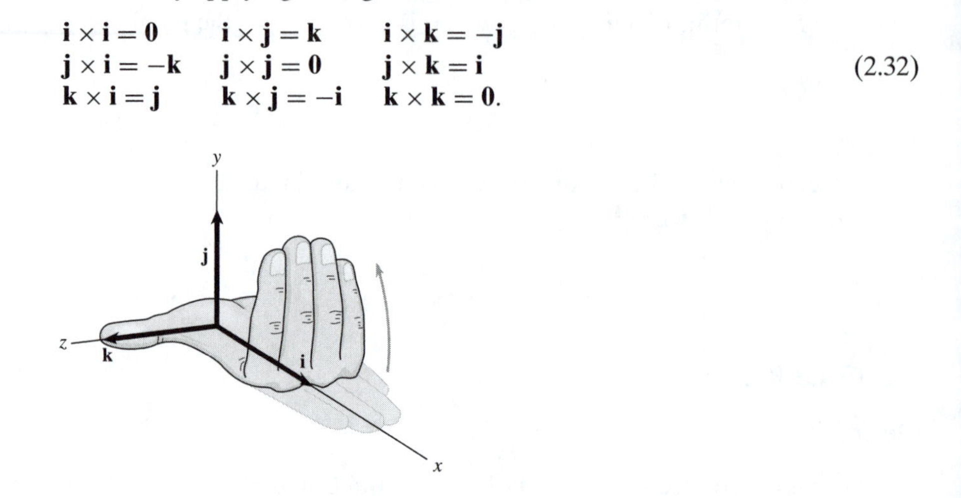

The right-hand rule indicates that $\mathbf{i} \times \mathbf{j} = \mathbf{k}$.

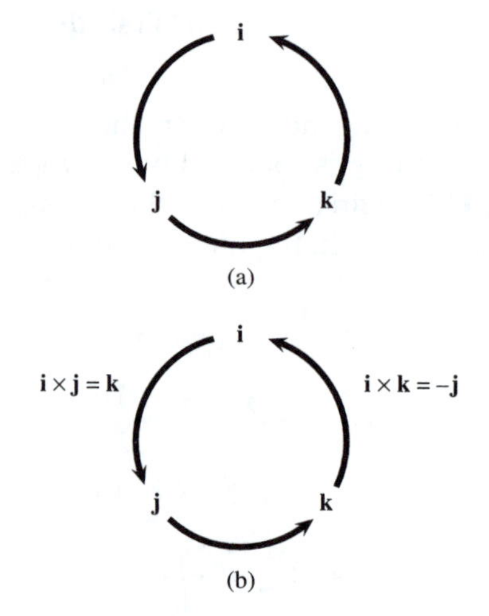

(a) Arrange the unit vectors in a circle with arrows to indicate their order. (b) You can use the circle to determine their cross products.

- In terms of scalar components

$$\mathbf{U} \times \mathbf{V} = \begin{vmatrix} \mathbf{i} & \mathbf{j} & \mathbf{k} \\ U_x & U_y & U_z \\ V_x & V_y & V_z \end{vmatrix}. \quad (2.34)$$

Mixed Triple Products

- The mixed triple product is the operation $\mathbf{U} \cdot (\mathbf{V} \times \mathbf{W})$.
- In terms of scalar components,

$$\mathbf{U} \cdot (\mathbf{V} \times \mathbf{W}) = \begin{vmatrix} U_x & U_y & U_z \\ V_x & V_y & V_z \\ W_x & W_y & W_z \end{vmatrix}. \quad (2.36)$$

- Interchanging any two of the vectors in the mixed triple product changes the sign, but not the absolute value, of the result. For example

$$\mathbf{U}\cdot(\mathbf{V}\times\mathbf{W}) = -\mathbf{W}\cdot(\mathbf{V}\times\mathbf{U})\,.$$

- If the vectors **U**, **V**, and **W** in the figure below form a right-handed system, it can be shown that the volume of the parallelepiped equals $\mathbf{U}\cdot(\mathbf{V}\times\mathbf{W})$.

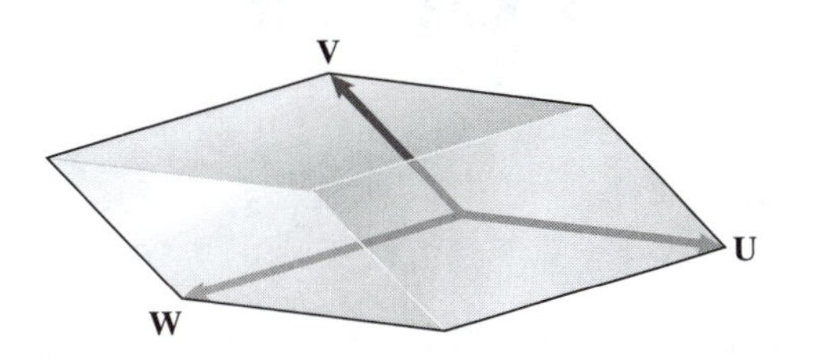

Parallelepiped defined by the vectors **U**, **V**, and **W**.

Helpful Tips and Suggestions

- Be aware of the differences between vectors and scalars. For example, force, velocity, and acceleration are *vectors* while speed, time and distance are *scalars*. If you are asked to find a vector (e.g., a force) you must report *both* magnitude and direction.
- Vector operations are essential in describing the basic principles of mechanics. Make sure you take the time to review basic vector algebra. It doesn't take long but the payoff (in terms of your effectiveness in mechanics) is significant.

Review Questions

1. How are the scalar components of a two-dimensional vector defined in terms of a Cartesian coordinate system?
2. If you know the scalar components of a two-dimensional vector, how can you determine its magnitude?
3. Suppose you know the coordinates of two points A and B. How do you determine the scalar components of the position vector of point B relative to point A?
4. How do you identify a right-handed coordinate system?
5. What are the direction cosines of a vector? If you know them, how do you determine the components of the vector?
6. What is the definition of the dot product? Is the dot product a vector or a scalar?
7. If the dot product of two vectors is zero, what does that mean?
8. If you know the components of two vectors **U** and **V**, how can you determine their dot product?
9. If **U**, **V**, and **W** are vectors, simplify
   (i) $\mathbf{U}\cdot(\mathbf{V}+\mathbf{W})$
   (ii) $(a\mathbf{V})\cdot\mathbf{U}$ where a is a scalar.
10. How can you use the dot product to determine the components of a vector parallel and normal to a straight line?

3

Forces

Main Goals of this Chapter:

In this chapter, we:

- Discuss the forces that occur frequently in engineering applications.
- Introduce the concept of the *free-body diagram* (used to identify forces on objects).
- Introduce the concept of *equilibrium* (used to determine unknown forces).

3.1 Forces, Equilibrium and Free-Body Diagrams

Terminology

- **Line of Action.** When a force is represented by a vector, the straight line collinear with the vector is called the *line of action* of the force.

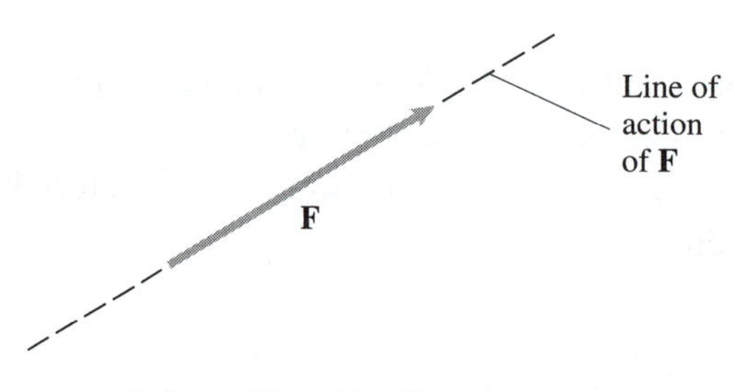

A force **F** and its line of action.

- **Systems of Forces.** A system of forces is *coplanar* or *two-dimensional* if the lines of action of the forces lie in the plane. Otherwise it is *three-dimensional.* A system of forces is *concurrent* if the lines of action of the forces intersect at a point and *parallel* if the lines of action are parallel.

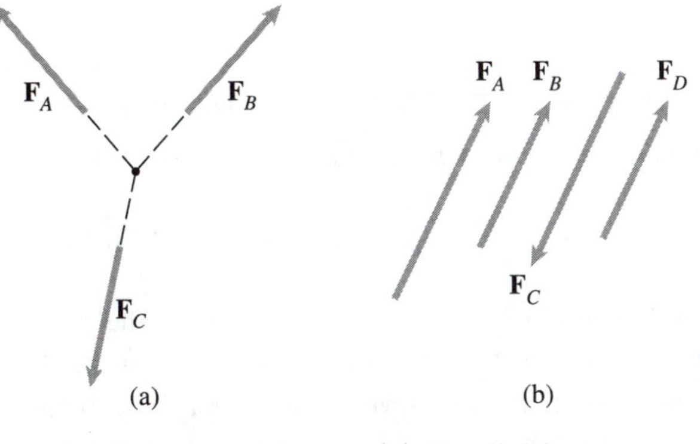

(a) Concurrent forces. (b) Parallel forces.

- **External and Internal Forces.** We say that a given object is subjected to an *external* force if the force is exerted by a different object. When one part of a given object is subjected to a force by another part of the same object, we say it is subjected to an *internal* force.
- **Body and Surface Forces.** A force acting on an object is called a *body force* if it acts on the volume of the object and a *surface force* if it acts on its surface.

Gravitational Forces

- The *gravitational force* or *weight* of an object has magnitude $|\mathbf{W}| = mg$ where g is the acceleration due to gravity at sea level ($g = 2.81\ m/s^2$ or $g = 32.2\ ft/s^2$). Gravitational forces (as well as electromagnetic forces) *act at a distance.*

Contact Forces

Contact forces are the forces that result from contacts between objects:

- **Surfaces.** Two surfaces in contact exert forces on each other that are equal in magnitude and opposite in direction. Each force can be resolved into the *normal force* and the *friction force*. If the friction force is negligible in comparison to the normal force, the surfaces are said to be *smooth*. Otherwise, they are *rough*.
- **Ropes and Cables.** A rope or cable attached to an object exerts a force on the object whose magnitude is equal to the tension and whose line of action is parallel to the rope or cable at the point of attachment. A pulley is a wheel with a grooved rim that can be used to change the direction of a rope or cable. When a pulley can turn freely and the rope or cable either is stationary or turns the pulley at a constant rate, the tension is approximately the same on both sides of the pulley.
- **Springs.** The force exerted by a linear spring is

$$|\mathbf{F}| = k|L - L_0|, \tag{3.2}$$

where k is the spring constant, L is the length of the spring, and L_0 is its unstretched length.

Equilibrium

- If an object is in *equilibrium*, the sum of the external forces acting on it is zero

$$\sum \mathbf{F} = \mathbf{0}. \tag{3.3}$$

In other words,

An object is in equilibrium only if each point of the object has the same constant velocity (steady translation).

The velocity must be measured relative to a frame of reference in which Newton's laws are valid (inertial frame).

To apply the equation of equilibrium (3.3), we must account for all the known and unknown forces ($\sum \mathbf{F}$) which act on the object. The easiest way to do this is to draw a *free-body diagram.*

- A free-body diagram is simply a sketch which shows the object 'free' from its surroundings with *all* the forces that act *on* it. There are three main steps:
  - ♦ **Identify the object you want to isolate.** Your choice is often dictated by particular forces you want to determine.
  - ♦ **Draw a sketch of the object isolated from its surroundings, and show relative dimensions and angles.** Imagine the object to be isolated or cut 'free' from its surroundings by drawing its outlined shape. Your drawing should be reasonably accurate, but it can omit irrelevant details.
  - ♦ **Draw vectors representing all of the external forces acting on the isolated object, and label them.** Don't forget to include the gravitational force if you are not intentionally neglecting it. The forces which are known should be labeled with their proper magnitudes and directions. Letters are used to represent the magnitudes and directions of forces that are unknown.
- **Coordinate System.** You will also need to choose a coordinate system so that you can express the forces on the isolated object in terms of components. Often you will find it convenient to choose the coordinate system before drawing the free-body diagram, but in some situations the best choice will not be apparent until after you have drawn the diagram.
- **Examples.** There are several examples and practice problems, as well as much more on drawing free-body diagrams in Part I of this study guide.

3.2 Two-Dimensional Force Systems

- Suppose that the system of external forces acting on an object in equilibrium is two-dimensional (coplanar). By orienting a coordinate system so that the forces lie in the $x-y$ plane, we can write the equilibrium equation (3.3) in scalar component form as

$$\sum F_x = 0, \quad \sum F_y = 0. \tag{3.4}$$

In other words,

The sums of the x and y *components of the external forces acting on an object in equilibrium must each equal zero.*

3.3 Three-Dimensional Force Systems

- Suppose that the system of external forces acting on an object in equilibrium is three-dimensional. We can write the equilibrium equation (3.1) in scalar component form as

$$\sum F_x = 0, \quad \sum F_y = 0, \quad \sum F_z = 0. \tag{3.5}$$

In other words,

The sums of the x, y and *z components of the external forces acting on an object in equilibrium must each equal zero.*

Helpful Tips and Suggestions

- Since we must account for *all the forces acting on the object* when applying the equations of equilibrium, the importance of *first* drawing a free-body diagram cannot be over-emphasized.
- **One of the most common mistakes made in writing equilibrium conditions is forgetting to include all of the forces acting.** When drawn carefully, a free-body diagram will make it easier for you to identify *all* the forces acting.
- Use Part I of this supplement to get lots of practice in drawing free-body diagrams and applying the equations of equilibrium for a particle.

Review Questions

1. What's the difference between a two- and three-dimensional system of forces?
2. What does it mean when a surface is *smooth*?
3. What is meant by 'equilibrium of an object'?
4. What do you know about the sum of the external forces acting on an object in equilibrium?
5. What are the three steps in drawing a free-body diagram?
6. What is a coplanar force system?
7. What is a three-dimensional system of forces?
8. What is the difference between equilibrium of coplanar and three-dimensional force systems?
9. What is the relation between the magnitude of the force exerted on a linear spring and the change in its length?
10. The following is the correct free-body diagram for the upper block. True or False?

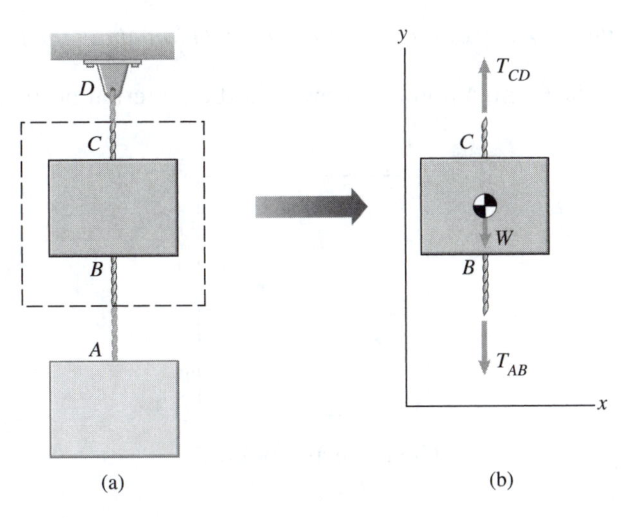

(a) Isolating the upper block to determine the tension in cable CD. (b) Free-body diagram of the upper block.

4

Systems of Forces and Moments

Main Goals of this Chapter:

- To discuss the concept of the *moment of a force* about a point and about a line and explain how to evaluate them.
- To introduce the concept of a couple.
- To define equivalent system of forces and moments.

4.1 Two-Dimensional Description of the Moment

The moment of a force about a point is the measure

of the tendency of the force to cause rotation about the point.

Consider a force of magnitude F and a point P, viewed in the direction perpendicular to the plane containing both.

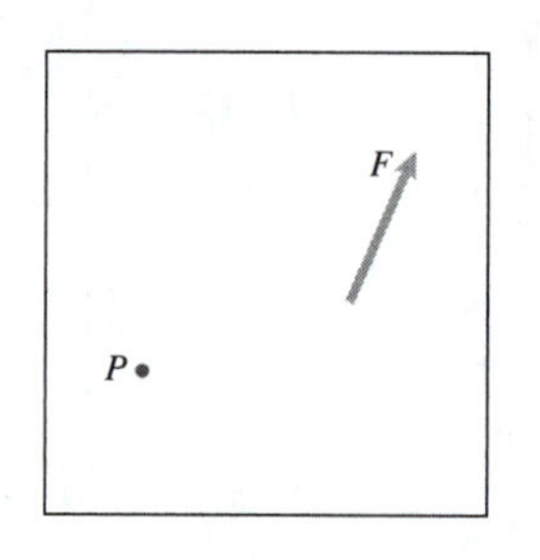

The force and point P.

- The *magnitude of the moment* of the force *about the point* P is

$$M_P = DF. \tag{4.1}$$

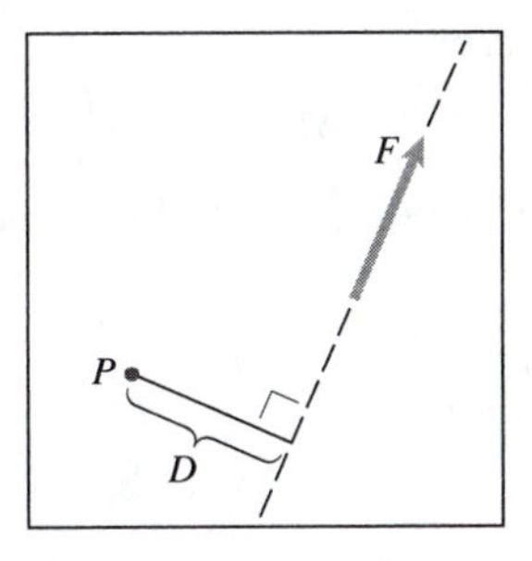

The perpendicular distance D from point P to the line of action of F.

- The *sense of the moment* of the force *about the point* P is counterclockwise.

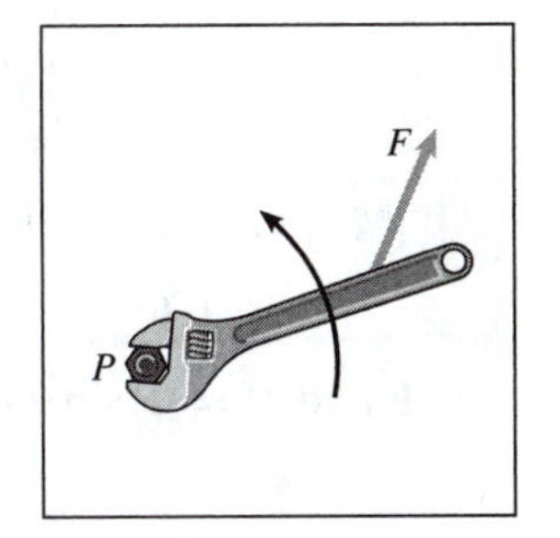

The sense of the moment is counterclockwise.

In general, we define counterclockwise moments to be positive and clockwise moments to be negative.

- The *dimensions of the moment* are (distance) $\times$ (force).

Note: *If the line of action of F passes through P, the perpendicular distance $D = 0$ and the moment of F about P is zero.*

The method described in this section can be used to determine the sum of the moments of a system of forces *about a point* if the forces are two-dimensional (coplanar) and the point lies in the same plane.

4.2 The Moment Vector

Consider a force vector **F** and a point P.

The force F and point P.

- The moment of **F** about point P is the *vector*

$$\mathbf{M}_P = \mathbf{r} \times \mathbf{F}, \tag{4.2}$$

where **r** is the position vector from P to *any* point on the line of action of **F**.

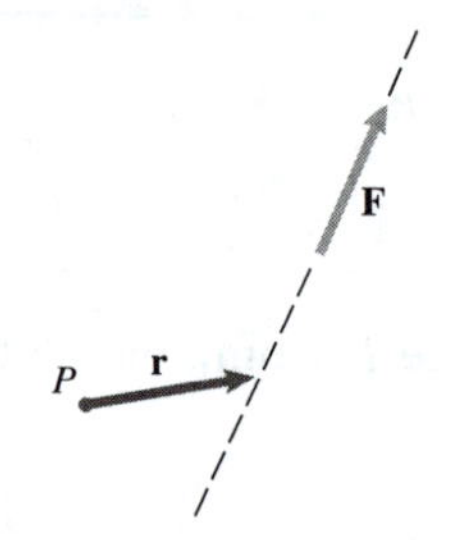

A vector r from P to a point on the line of action of **F**.

Magnitude of the Moment

- The *magnitude* of vector $\mathbf{M}_P$ is given by

$$\begin{aligned} |\mathbf{M}_P| &= |\mathbf{r}|\,|\mathbf{F}| \sin\theta \\ &= D\,|\mathbf{F}|, \end{aligned} \tag{4.3}$$

where θ is the angle between the tails of **r** and **F** and $D = |\mathbf{r}| \sin\theta$ is the perpendicular distance from the point P to the line of action of **F**.

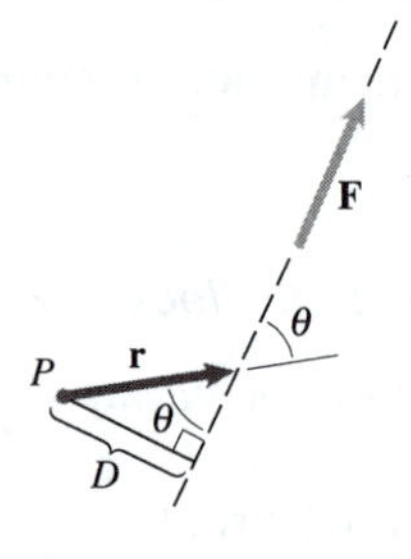

The angle θ and the perpendicular distance D.

Direction of the Moment

- The vector $\mathbf{M}_P$ is perpendicular to the plane containing P and **F**.

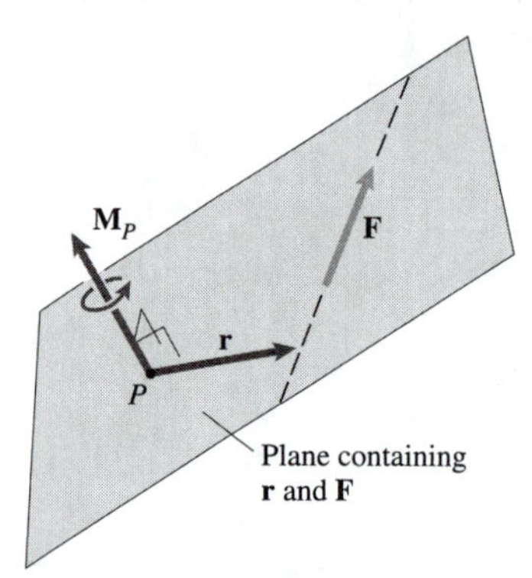

$\mathbf{M}_P$ is perpendicular to the plane containing p and **F**.

- The direction of $\mathbf{M}_P$ indicates the *sense of the moment*:

  If you point the thumb of your right hand in the direction of $\mathbf{M}_P$, *the 'arc' of your fingers indicates the sense of the rotation that* $\mathbf{F}$ *tends to cause about* P *(Right-hand Rule!).*

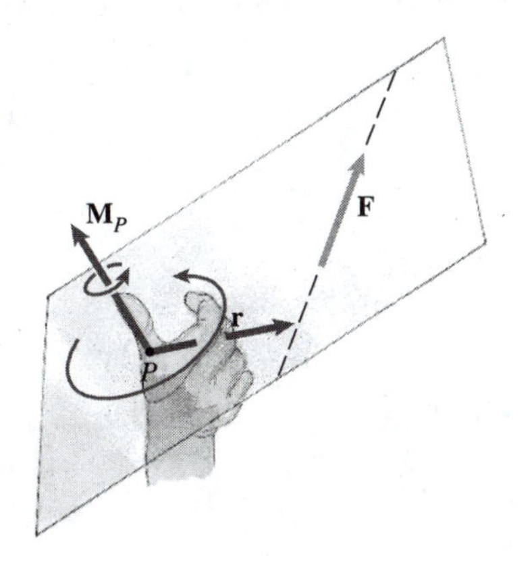

The direction of $\mathbf{M}_P$ indicates the sense of the moment.

Relation to the Two-Dimensional Description

If our view is perpendicular to the plane containing point P and the force $\mathbf{F}$, the two-dimensional description of the moment used in Section 4.1 specifies both the magnitude and direction of $\mathbf{M}_P$. In this situation, $\mathbf{M}_P$ is perpendicular to the page, and the right-hand rule indicates whether it points out of or into the page.

Varignon's Theorem

Let $\mathbf{F}_1, \mathbf{F}_2, \ldots \mathbf{F}_N$ be a concurrent system of forces whose lines of action intersect at point Q.

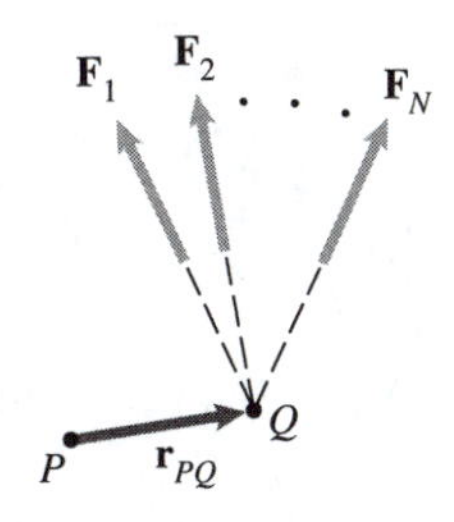

A system of concurrent forces and a point P.

The moment of the system about point P is

$$(\mathbf{r}_{PQ} \times \mathbf{F}_1) + (\mathbf{r}_{PQ} \times \mathbf{F}_2) + \ldots + (\mathbf{r}_{PQ} \times \mathbf{F}_N)$$
$$= \mathbf{r}_{PQ} \times (\mathbf{F}_1 + \mathbf{F}_2 + \ldots + \mathbf{F}_N),$$

where $\mathbf{r}_{PQ}$ is the vector from P to Q. In other words:

The moment of a force about a point P *is equal to the sum of the moments of its components about* P.

4.3 Moment of a Force About a Line

Definition

- *The moment of a force about a line is the measure of the tendency of the force to cause rotation about the line.*
- Consider a line L and force $\mathbf{F}$. Let $\mathbf{M}_P$ be the moment of $\mathbf{F}$ about an arbitrary point P on L. *The moment of* $\mathbf{F}$ *about* L *is the component of* $\mathbf{M}_P$ *parallel to* L*, which we denote by* $\mathbf{M}_L$.

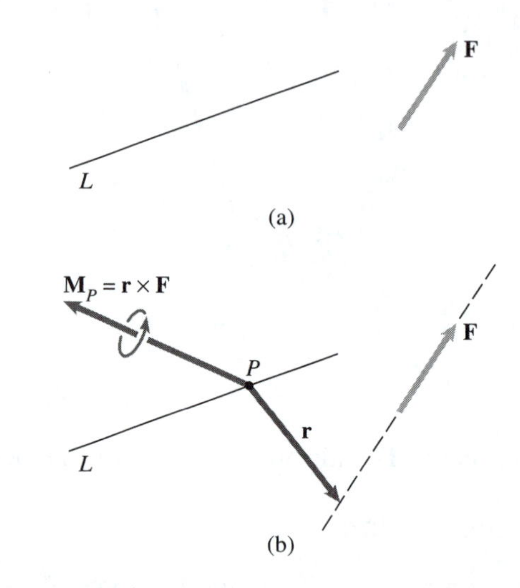

(a) The line L and force $\mathbf{F}$. (b) $\mathbf{M}_P$ is the moment of $\mathbf{F}$ about any point P on L.

- The *magnitude* of the moment of $\mathbf{F}$ about L is $|\mathbf{M}_L|$, and when the thumb of the right hand is pointed in the direction of $\mathbf{M}_L$, the arc of the fingers indicates the sense of the moment about L.

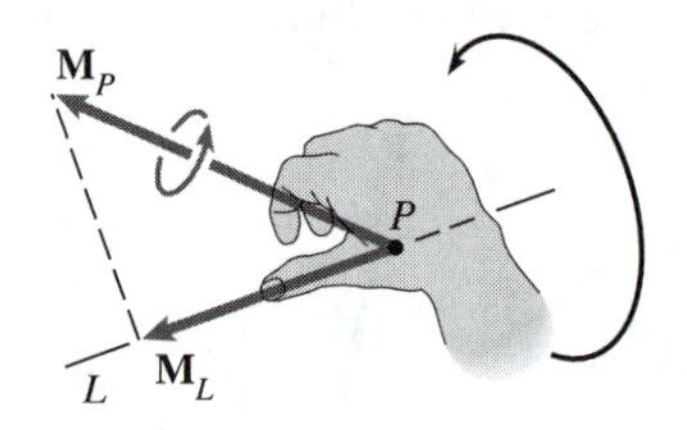

The component $\mathbf{M}_L$ is the moment of $\mathbf{F}$ about L.

- In terms of a unit vector $\mathbf{e}$ along L,

$$\mathbf{M}_L = (\mathbf{e}.\mathbf{M}_P)\,\mathbf{e} \tag{4.4}$$

$$= [\mathbf{e}.(\mathbf{r} \times \mathbf{F})]\,\mathbf{e} \tag{4.5}$$

♦ When the line of action of $\mathbf{F}$ is perpendicular to a plane containing L, the magnitude of the moment of $\mathbf{F}$ about L is equal to the product of the magnitude of $\mathbf{F}$ and the perpendicular distance D from L to the point where the line of action intersects the plane i.e.,

$$|\mathbf{M}_L| = |\mathbf{F}|\,D.$$

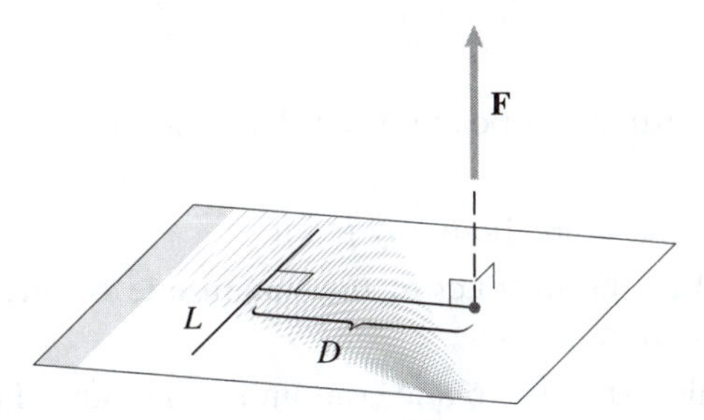

F is perpendicular to a plane containing L.

♦ When the line of action of **F** is parallel to L, the moment of **F** about L is zero: $\mathbf{M}_L = \mathbf{0}$. Since $\mathbf{M}_P = \mathbf{r} \times \mathbf{F}$ is perpendicular to **F**, $\mathbf{M}_P$ is perpendicular to L and the vector component of $\mathbf{M}_P$ parallel to L is zero.

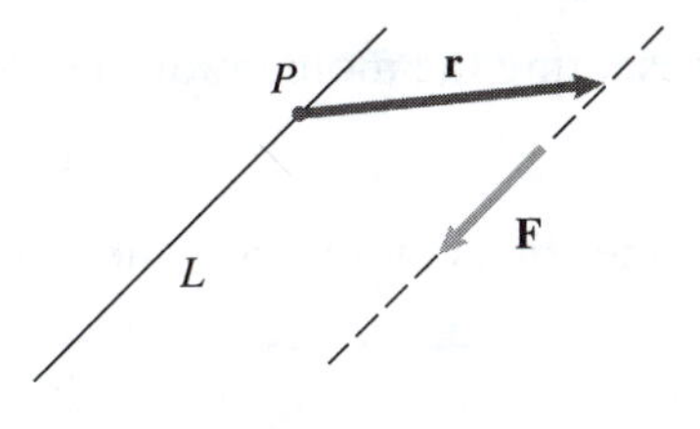

F is parallel to L.

♦ When the line of action of **F** intersects L, the moment of **F** about L is zero. Since we can choose any point on L to evaluate $\mathbf{M}_P$, we can use the point where the line of action of **F** intersects L. The moment $\mathbf{M}_P$ about that point is zero, so its vector component parallel to L is zero.

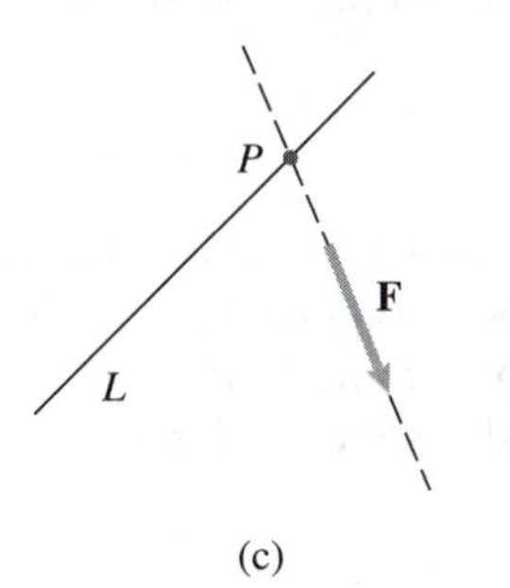

(c)

The line of action of **F** intersects L at P.

Summary - Procedure for Determining the Moment of a Force About a Line

1. Determine a vector **r**—Choose any point P on L and determine the components of vector **r** from P to any point on the line of action of **F**.
2. Determine a vector **e**—Determine the components of a unit vector along L. It doesn't matter in which direction along L it points.
3. Evaluate $\mathbf{M}_L$—You can calculate $\mathbf{M}_P = \mathbf{r} \times \mathbf{F}$ and determine $\mathbf{M}_L$ using the equation

$$\mathbf{M}_L = (\mathbf{e}.\mathbf{M}_P)\,\mathbf{e} = [\mathbf{e}.\,(\mathbf{r} \times \mathbf{F})]\,\mathbf{e}.$$

4.4 Couples

- Two forces that have equal magnitudes, opposite directions, and do not have the same line of action are called a *couple*.
    - ♦ The moment **M** of a couple is the same about any point.
    - ♦ The *magnitude* of **M** is equal to the product of the magnitude of one of the forces and the perpendicular distance between the lines of action.
    - ♦ The *direction* of **M** is perpendicular to the plane containing the lines of action.
- Because a couple exerts a moment but *no net force*, it can be represented by showing the moment vector.

Representing the moment of the couple.

OR, the couple can be represented in two dimensions by showing the magnitude of the moment and a circular arrow to indicate the sense

The moment represented in this way is called the *moment of a couple* or simply a *couple.*

4.5 Equivalent Systems

A system of forces and moments is simply a *particular set* of forces and moments of couples. The systems of forces and moments dealt with in engineering can be complicated. This is especially true in the case of distributed forces, such as the pressure forces exerted by water on a dam. Fortunately, if we are concerned only with the *total force and moment exerted,* we can represent complicated systems of forces and moments by much *simpler* systems.

- Two systems of forces and moments are defined to be equivalent if the sums of the forces are equal

$$\left(\sum \mathbf{F}\right)_1 = \left(\sum \mathbf{F}\right)_2 \tag{4.7}$$

and the sums of the moments about a point P are equal,

$$\left(\sum \mathbf{M}_P\right)_1 = \left(\sum \mathbf{M}_P\right)_2 . \tag{4.8}$$

- *If the sums of the forces are equal for two systems of forces and moments and the sums of the moments about one point P are equal, then the sums of the moments about any point are equal.*

Representing Systems by Equivalent Systems

- If the system of forces and moments acting on an object is represented by an equivalent system, the equivalent system exerts the same total force and total moment on the object.
- *Any* system can be represented by an equivalent system consisting of a force **F** acting at a given point P and a couple **M**.

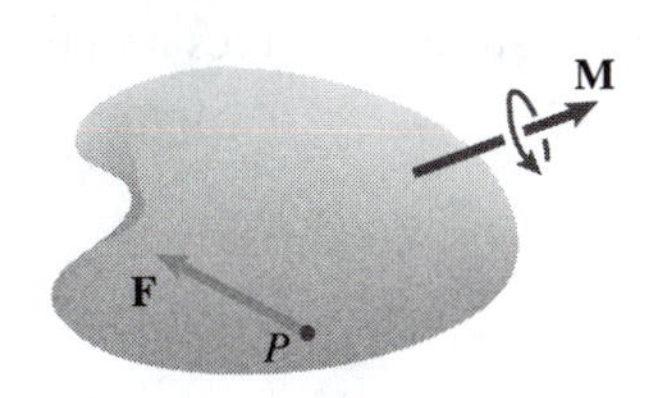

A force acting at P and a couple.

- The simplest system that can be equivalent to any system of forces and moments is the *wrench*, which is a force **F** and a couple $\mathbf{M}_P$ that is parallel to **F**.

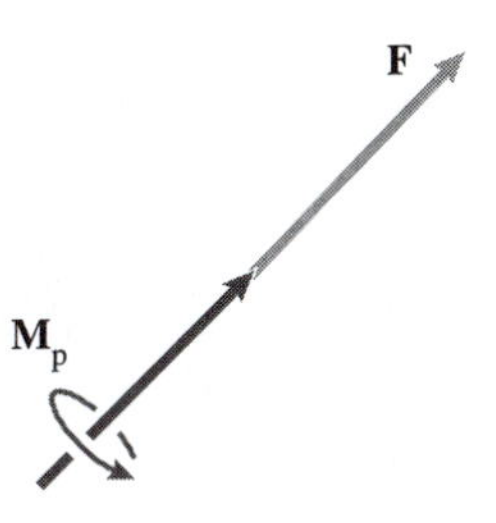

- A system of *concurrent* forces can be represented by a *single force*.
- A system of *parallel* forces whose sum is not zero can be represented by a *single force*.

Helpful Tips and Suggestions

- Be aware of the difference between the *cross product* and the *dot product* of two vectors. The former is a *vector* while the latter is a *scalar*. Also

$$\mathbf{A} \cdot \mathbf{B} = \mathbf{B} \cdot \mathbf{A} \quad \textbf{BUT} \quad \mathbf{A} \times \mathbf{B} \neq \mathbf{B} \times \mathbf{A} \ (= -\mathbf{A} \times \mathbf{B}).$$

- The right-hand rule is essential in the calculation of moments. You should be able to apply this rule quickly and accurately.

Review Questions

1. What is meant by a *moment*? Is it a vector or a scalar?
2. What's the magnitude of the moment of a force about a point?
3. How do you calculate the sense (direction) of a moment about some point P?
4. If the line of action of a force passes through a point P, what do you know about the moment of the force about P?
5. When you use the equation $\mathbf{M}_P = \mathbf{r} \times \mathbf{F}$ to determine the moment of a force **F** about P, how do you choose **r**?

6. If you know the components of the vector $\mathbf{M}_P = \mathbf{r} \times \mathbf{F}$, how can you determine the product of the magnitude of $\mathbf{F}$ and the perpendicular distance D from P to the line of action of $\mathbf{F}$?
7. How do you figure out the sense of the moment of $\mathbf{F}$ about P using the formula $\mathbf{M}_P = \mathbf{r} \times \mathbf{F}$?
8. How would you calculate the moment exerted about a point P by a couple consisting of forces $\mathbf{F}$ and $-\mathbf{F}$?
9. True or false? The moment of a couple about O is different than the moment of the same couple about $P \neq O$?
10. What is meant by an *equivalent system*?
11. What conditions must be satisfied for two systems of forces and moments to be *equivalent*?
12. Define what is meant by a *wrench*.

5

Objects in Equilibrium

Main Goals of this Chapter:

- To use the equilibrium equations to analyze the forces and couples acting on many types of objects.
- To define the support conventions commonly used in engineering.
- To discuss situations that can result in an object's being *statically indeterminate.*
- To define two-force and three-force members.

5.1 Two-Dimensional Applications

Background

- When an object acted upon by a system of forces and moments is in equilibrium, the following conditions are satisfied.

$$\sum \mathbf{F} = \mathbf{0}, \tag{5.1}$$

$$\sum \mathbf{M}_{(any\ point)} = \mathbf{0}. \tag{5.2}$$

The Scalar Equilibrium Equations

- When the loads and reactions on an object in equilibrium form a two-dimensional system of forces and moments, they are related by three scalar equilibrium equations:

$$\sum F_x = 0, \tag{5.3}$$

$$\sum F_y = 0, \tag{5.4}$$

$$\sum M_{(anypoint)} = 0. \tag{5.5}$$

No more than three equilibrium equations can be obtained from a given two-dimensional free-body diagram. This means that we can solve for at most three unknown forces or couples.

Supports

- Forces and couples exerted on an object by its supports are called *reactions*. The other forces and couples are the *loads*.
- Common supports are represented by models called *support conventions*. Actual supports often closely resemble the support conventions, but even when they don't, we represent them by these conventions if the actual supports exert the same (or approximately the same) reactions as the models.
- The different types of supports arising in two-dimensional applications, together with their corresponding reactions, are listed in Table 2.1 in Part I of this supplement (or Table 5.1 of the text).

Free-Body Diagrams

- No equilibrium problem should be solved without *first* drawing the free-body diagram, so as to account for *all* the forces *and couple moments* that act on the object. By using the support conventions, we can model more elaborate objects and construct their free-body diagrams in a systematic way.
- Part I of this supplement is devoted to the drawing of free-body diagrams including, specifically, *free-body diagrams for objects in two dimensional applications*. Study Part I of this supplement making special note of the following important points:
  - If a support *prevents translation* of an object in a particular direction, then the support exerts a *force* on the object in that direction.
  - If *rotation is prevented*, then the support exerts a *couple moment* on the object.
  - Study Table 2.1 in Part I of this supplement (or Table 5.1 of the text) where support conventions commonly used in two-dimensional applications are summarized.
  - Internal forces are never shown on the free-body diagram since they occur in equal but opposite collinear pairs and therefore cancel out.
  - The weight of a body is an external force and its effect is shown as a single resultant force acting through the body's center of gravity G.
  - *Couple moments* can be placed anywhere on the free-body diagram since they are *free vectors*. Forces can act at any point along their lines of action since they are *sliding vectors*.

5.2 Statically Indeterminate Objects

There are two common situations in which the use of the equilibrium equations to determine unknown forces and couples acting on objects in equilibrium *does not lead to a solution:*

1. **More Unknowns than Equations.** The free-body diagram of an object can have more unknown forces or couples than the number of independent equilibrium equations you can obtain. Since you can write no more than *three* such equations for a given free-body diagram in a two-dimensional problem, when there are *more than three* unknowns, you cannot determine them from the equilibrium equations alone. For example:
   - **Redundant Supports**—When an object has more supports than the minimum number necessary to maintain it in equilibrium the object is said to have *redundant supports*.
2. **Inability to Maintain Equilibrium.** The second situation in which the use of the equilibrium equations to determine unknown forces and couples acting on objects in equilibrium *does not lead to a solution* arises when

the supports of an object are improperly designed such that they cannot maintain equilibrium under the loads acting on it. In this case the object is said to have *improper supports*.

In either situation, the object is *statically indeterminate*.

- The difference between the number of reactions and the number of independent equilibrium equations is called the *degree of redundancy*.
- Even if an object is statically indeterminate due to redundant supports, it may be possible to determine *some* of the reactions from the equilibrium equations.

5.3 Three-Dimensional Applications

The first step in solving three-dimensional equilibrium problems, as in the case of two dimensions, is to *draw a free-body diagram*. The general procedure for doing this is the same as that outlined for the two-dimensional case in Section 5.1 of the text. However, there are a few subtle differences of which you should be aware:

- It is necessary to be familiar with the different types of reactive forces and couple moments acting at various types of supports and connections when members are viewed in three dimensions. It is important to recognize the symbols used to represent each of these supports and to understand clearly how the forces and couple moments are developed by each support. These are summarized in Table 5-2 of the text. Remember:
  — *As in the two-dimensional case, a force is developed by a support that restricts the translation of the attached member, whereas a couple moment is developed when rotation of the attached member is prevented.*

The Scalar Equilibrium Equations

- In the case of a three-dimensional system of forces and moments, you can obtain up to six independent equilibrium equations: the three components of the sum of the forces must equal zero, and the three components of the sum of moments about any point must equal zero.

- The loads and reactions on an object in equilibrium satisfy the six scalar equilibrium equations:

$$\sum F_x = 0, \quad \sum F_y = 0, \quad \sum F_z = 0,$$
$$\sum M_x = 0, \quad \sum M_y = 0, \quad \sum M_z = 0.$$

  You can evaluate the sums of the moments about any point.

More than six independent equilibrium equations cannot be obtained from a given free-body diagram so we can solve for at most six unknown forces or couples.

- The procedure for determining the reactions on objects subjected to three-dimensional systems of forces and moments (drawing the free-body diagram and applying the equilibrium equations) is the same as in two-dimensions. The only difference is the support conventions commonly used in three-dimensional applications. These are summarized in Table 5.2.

Table 5.2

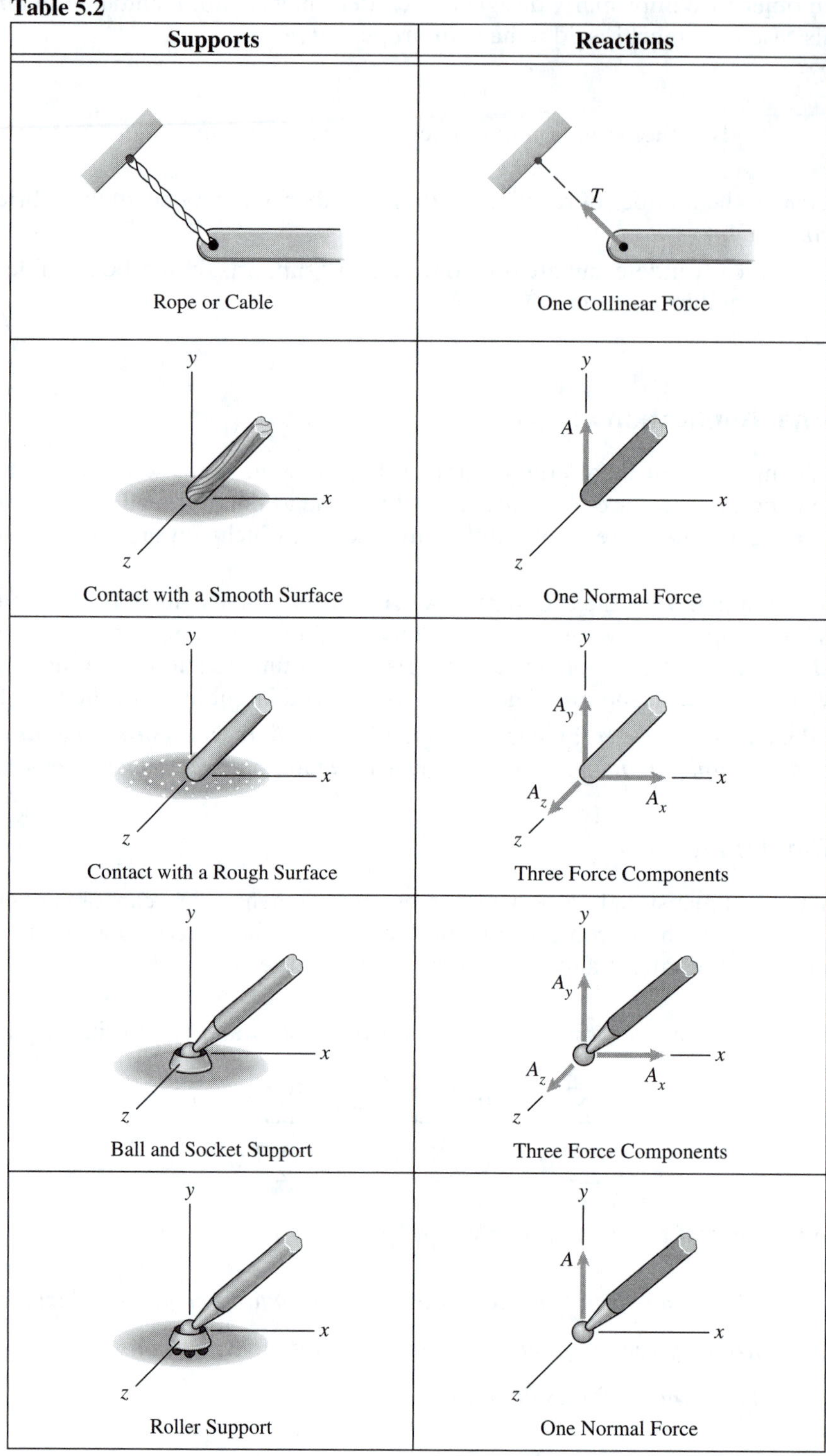

Supports	Reactions
Rope or Cable	One Collinear Force
Contact with a Smooth Surface	One Normal Force
Contact with a Rough Surface	Three Force Components
Ball and Socket Support	Three Force Components
Roller Support	One Normal Force

Table 5.2 *continued*

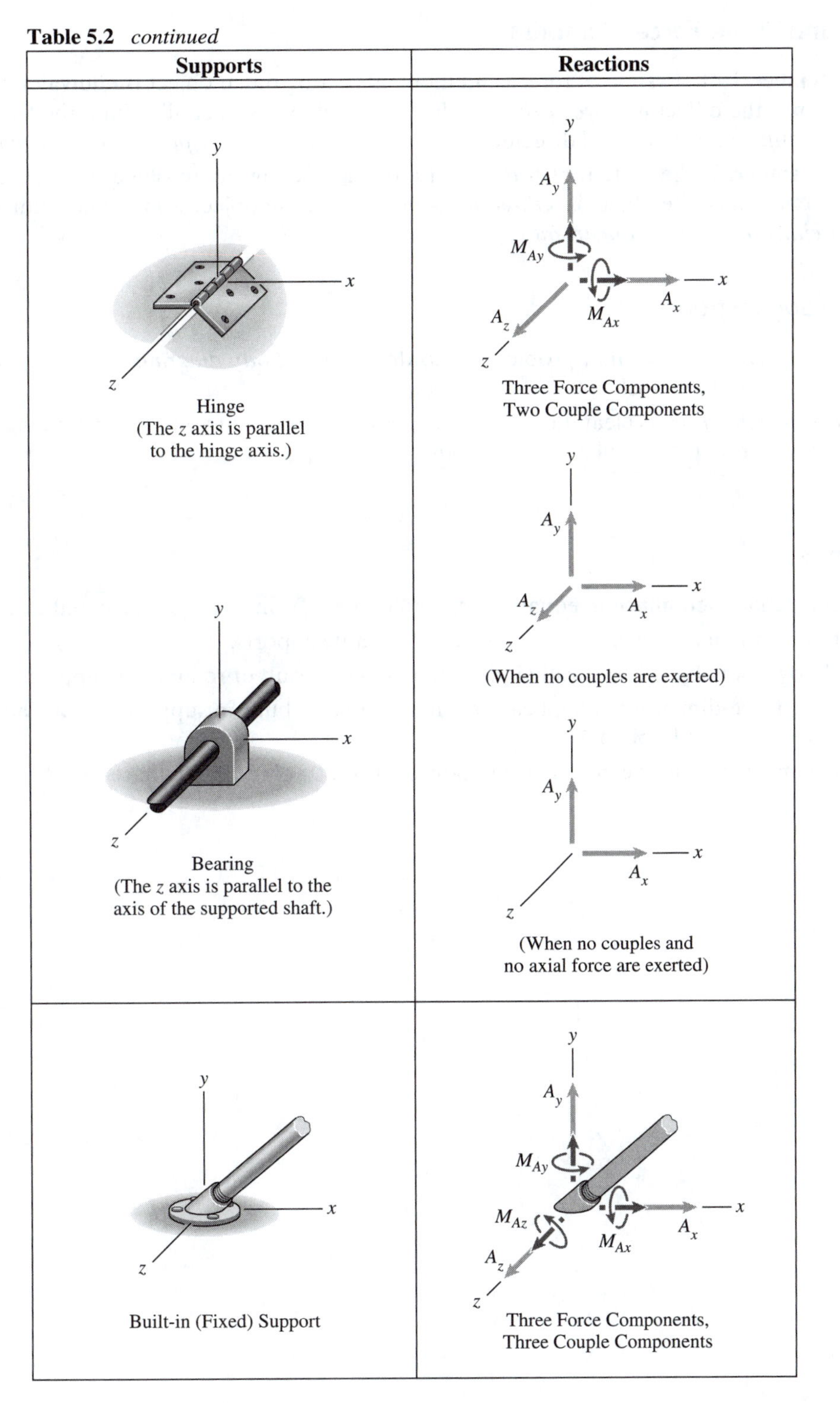

5.4 Two-Force and Three-Force Members

- **Two-Force Member.** If the system of forces and moments acting on an object is equivalent to two forces acting at different points, the object is a *two-force member.* If the object is in equilibrium, the two forces are *equal in magnitude, opposite in direction*, and directed *along the line through their points of application.*
- **Three-Force Member.** If the system of forces and moments acting on an object is equivalent to three forces acting at different points, the object is a *three-force member.* If the object is in equilibrium, the three forces are *coplanar and either parallel or concurrent.*

Helpful Tips and Suggestions

- The first step in solving equilibrium problems is to draw a *free-body diagram.* Don't try to skip this stage no matter how trivial you think it is!
- Make the *free-body diagram* as clear and concise as possible. It will aid your understanding of the problem and it will help you construct the equilibrium equations.

Review Questions

1. How many independent equilibrium equations can you obtain from a two-dimensional free-body diagram?
2. What does it mean when an object is said to have redundant supports.
3. How do you know if an object is statically indeterminate as a result of redundant supports?
4. If, in a particular three-dimensional application, an object has a built-in support and any additional supports, is it statically indeterminate? Explain why.
5. In general, how many reactions can a hinge support exert on an object? What are they?

6

Structures in Equilibrium

Main Goals of this Chapter:

- To analyze structures composed of interconnected parts or members:
  - ♦ *Trusses*: structures composed entirely of two-force members.
  - ♦ *Frames*: structures designed to remain stationary and support loads.
  - ♦ *Machines*: structures designed to move and exert loads.

6.1 Trusses

- *A structure of members interconnected at joints is a truss if it is composed entirely of two-force members.*
- A member of a truss is in *tension* if the axial forces at the ends are directed away from each other.
- A member of a truss is in *compression* if the axial forces at the ends are directed toward each other.
- The axial forces in the members of trusses are found using:
  - ♦ *The method of joints*: when you need to determine the axial forces in *all* members of a truss.
  - ♦ *The method of sections*: when you need to determine only the axial forces in a *few* members

6.2 The Method of Joints

- The *method of joints* involves drawing free-body diagrams of the joints of a truss one by one and using the equilibrium equations to determine the axial forces in the members. Before beginning it is usually necessary to draw a free-body diagram of the entire truss (i.e., treat the truss as a single object) and determine the reactions at its supports.
- In two-dimensions, you can obtain only *two* independent equilibrium equations from the fr ee-body diagram of a joint.
- When you determine the axial forces in the members of a truss, your task will often be simpler if you are familiar with three particular types of joints:
  - — **Truss joints with two collinear members and no load.** The sum of the forces must equal zero, $T_1 = T_2$. The axial forces are equal.

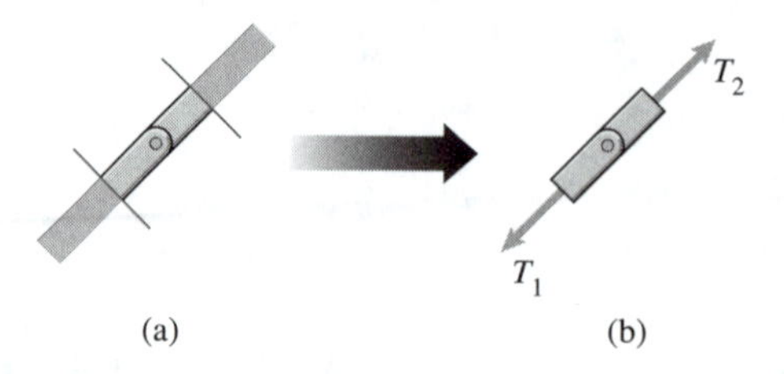

(a) A joint with two collinear members and no load. (b) Free-body diagram of the joint.

— **Truss joints with two noncollinear members and no load.** Because the sum of the forces in the x-direction must equal zero, $T_2 = 0$. Therefore T_1 must also equal zero. The axial forces are equal.

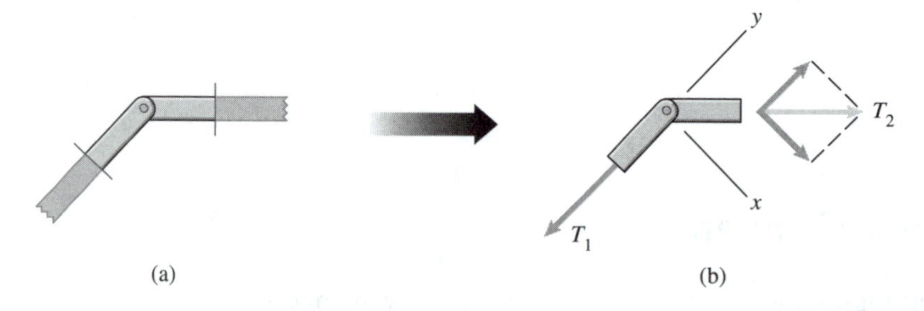

(a) A joint with two noncollinear members and no load. (b) Free-body diagram of the joint.

— **Truss joints with three members, two of which are collinear, and no load.** Because the sum of the forces in the x-direction must equal zero, $T_3 = 0$. The sum of the forces in the y-direction must equal zero so $T_1 = T_2$. The axial forces in the collinear members are equal, and the axial force in the third member is zero.

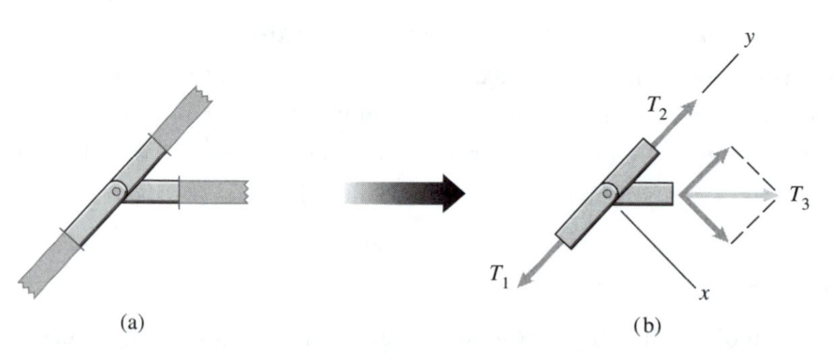

(a) A joint with three members, two of which are collinear, and no load. (b) Free-body diagram of the joint.

6.3 The Method of Sections

- When we need to know the axial forces *only in certain members of a truss*, we often can determine them more quickly using the *method of sections* than using the method of joints.
- The *method of sections* involves drawing free-body diagrams of parts, or sections, of a truss and using the equilibrium equations to determine the axial forces in selected members.

Comparing the Method of Sections to the Method of Joints

- *Both* methods involve cutting members to obtain free-body diagrams of parts of a truss.
- In the *method of joints*, we move from *joint to joint*, drawing free-body diagrams of the joints and determining the axial forces in the members as we go.
- In the *method of sections*, we try to obtain a *single* free-body diagram that allows us to determine the axial forces in *specific* members.
- In contrast to the free-body diagrams of joints, the forces on the free-body diagrams used in the *method of sections* are *not usually concurrent*, and we can obtain three independent equilibrium equations. Although there are exceptions, it is usually necessary to choose a section that requires cutting no more than *three* members, or there will be more unknown axial forces than equilibrium equations.

6.4 Space Trusses

- We can form a simple three-dimensional structure by connecting six bars at their ends to obtain a tetrahedron.

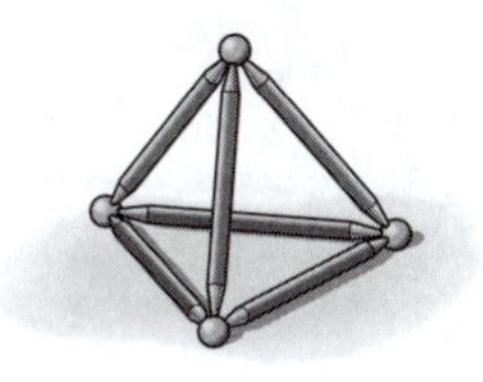

Space trusses with 6, 9, and 12 members.

- By adding members, we can obtain more elaborate structures.

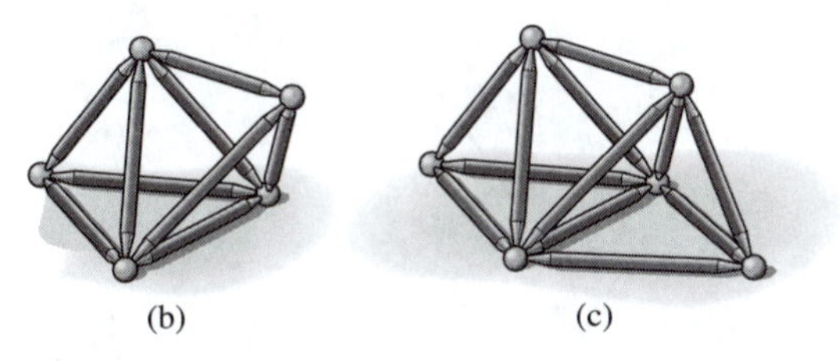

Space trusses with 6, 9, and 12 members.

- Three-dimensional structures such as these are called *space trusses* if they have joints that do not exert couples on the members (i.e., the joints behave like ball and socket supports (see Table 5.1)) and they are loaded and supported at their joints.
- Space trusses are analyzed by the same methods described for two-dimensional trusses. The only difference is the need to cope with the more complicated geometry.
- Three equilibrium equations can be obtained from the free-body diagram of a joint in three-dimensions, so it is usually necessary to choose joints to analyze that are subjected to known forces and *no more than* three unknown forces.

6.5 Frames and Machines

- Many structures such as the frame of a car and the human structure of bones, tendons, and muscles are not composed entirely of two-force members and thus cannot be modeled as trusses.

- Structures of interconnected members that *do not* satisfy the definition of a truss are called *frames* if they are designed to remain stationary and support loads and *machines* if they are designed to move and apply loads.
- When trusses are analyzed by cutting members to obtain free-body diagrams of joints or sections, the internal forces acting at the "cuts" are simple axial forces.

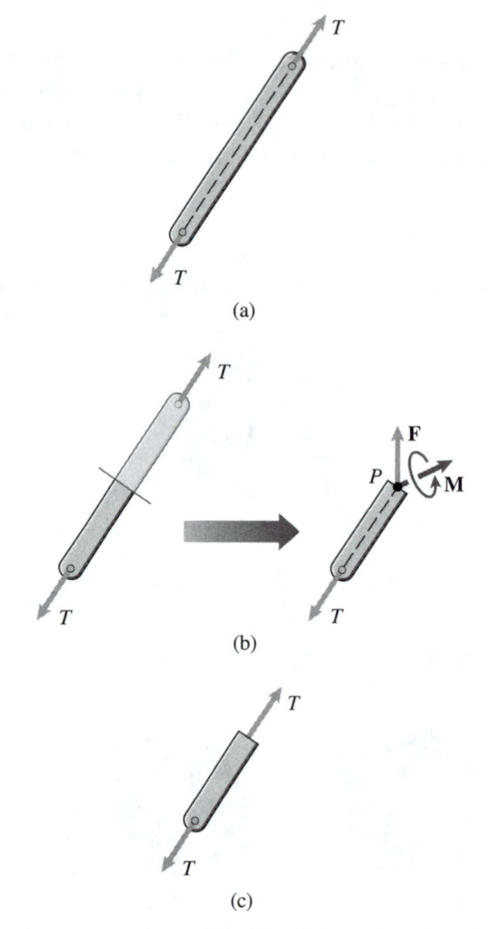

(a) Each member of a truss is a two-force member. (b) Obtaining the free-body diagram of part of the member. (c) The internal force is equal and opposite to the force acting at the joint, and the internal couple is zero.

This is not generally true for frames or machines and a different method of analysis is necessary. Instead of cutting members, you isolate entire members, or in some cases, combinations of members, from the structure.

Procedure for Determining Forces and Couples on the Members of a Frame or Machine

- **Free-Body Diagram**
  - Draw the free-body diagram of the entire structure (that is, treat the structure as a single object) and determine the reactions at its supports. *This step can greatly simplify your analysis of the members.* In some cases the entire structure will be statically indeterminate, but it is helpful to determine as many of the reactions as possible.
  - Next, draw free-body diagrams of individual members or selected combinations of members.

- **Equations of Equilibrium**
  - — Apply the equilibrium equations to determine the forces and couples acting on individual or selected combinations of members. *You can simplify this step by identifying two-force members.* This will reduce the number of unknown forces that must be determined.
  - — If a load acts at a joint of the structure, you can place the load on the free-body diagram of any one of the members attached at that joint.

Helpful Tips and Suggestions

- The importance of drawing and using a clear and concise free-body diagram cannot be overstated.
- As in most mechanics problems, *practice* is the key. Make sure you read Examples 6-1 through 6-8 in the text and attempt to draw the requested free-body diagrams *yourself.* When doing so, make sure the work is neat and that all the forces and couple moments are properly labelled.

Review Questions

1. What is a truss?
2. What is the method of joints?
3. In two dimensions, how many independent scalar equilibrium equations are available from the free-body diagram of a joint?
4. What is the method of sections?
5. What are the essential differences between the *method of sections* and the *method of joints*?
6. What methods are available to determine the forces developed in the members of a simple space truss?
7. What's the difference between a frame and a machine?
8. Why can't frames or machines be analyzed in the same way as trusses—by cutting members to obtain free-body diagrams of joints or sections?
9. What's the main difference in the methods of analysis for frames or machines as opposed to trusses?
10. What's the procedure for determining the forces and couples on the members of a frame or machine?

7

Centroids and Centers of Mass

Main Goals of this Chapter:

- To define *center of mass* and *centroid*.
- To show how to determine centroids of areas, composite areas, volumes and lines.
- To discuss the concept of a distributed load and how to determine the force and moment due to a distributed load.
- To use the theorems of Pappus and Guldinus for finding the area and volume of a surface of revolution.
- To show how to determine centers of mass of objects and composite objects.

7.1 Centroids of Areas

- A *centroid* is a weighted average position.
- The coordinates of the centroid of an area A in the $x-y$ plane are

$$\bar{x} = \frac{\int_A x dA}{\int_A dA}, \quad \bar{y} = \frac{\int_A x dA}{\int_A dA}, \qquad (7.6), (7.7)$$

  where x and y are the coordinates of the differential element of area dA. The subscript A on the integral signs means that the integration is carried out over the entire area.
- Keeping in mind that the centroid of an area is its average position will often help you locate it. For example,
  - ♦ The centroid of a circular area or a rectangular area obviously lies at the center of the area.
  - ♦ If an area has "mirror image" symmetry about an axis, the centroid lies on the axis.
  - ♦ If an area is symmetric about two axes, the centroid lies at their intersection.

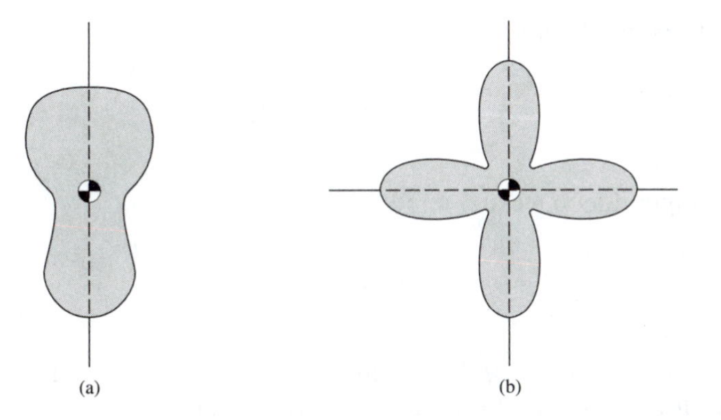

(a) An area that is symmetric about an axis. (b) An area with two axes of symmetry.

7.2 Centroids of Composite Areas

- The coordinates of the centroid of a composite area composed of an arbitrary number of parts $A_1, A_2, \ldots$ are

$$\bar{x} = \frac{\sum_i \bar{x}_i A_i}{\sum_i A_i}, \quad \bar{y} = \frac{\sum_i \bar{y}_i A_i}{\sum_i A_i}. \tag{7.9}$$

- When you can divide an area into parts whose centroids are known, you can use these expressions to determine its centroid. The centroids of some simple areas are tabulated in Appendix B in the text.

Procedure for Determining the Centroid of a Composite Area

1. *Choose the parts*—Try to divide the composite area into parts whose centroids you know or can easily determine.
2. *Determine the values for the parts*—Determine the centroid and the area of each part. Watch for instances of symmetry that can simplify your task.
3. *Calculate the centroid*—Use Equations (7.9) to determine the centroid of the composite area.

- Look at Examples 7.3–7.4 in the text which illustrate the procedure.

7.3 Distributed Loads

- A force distributed along a line is described by a function w $\left(\text{units of } \frac{\textit{force}}{\textit{length}}\right)$, defined such that the force on a differential element dx of the line is $w dx$. The graph of w is called the *loading curve*.

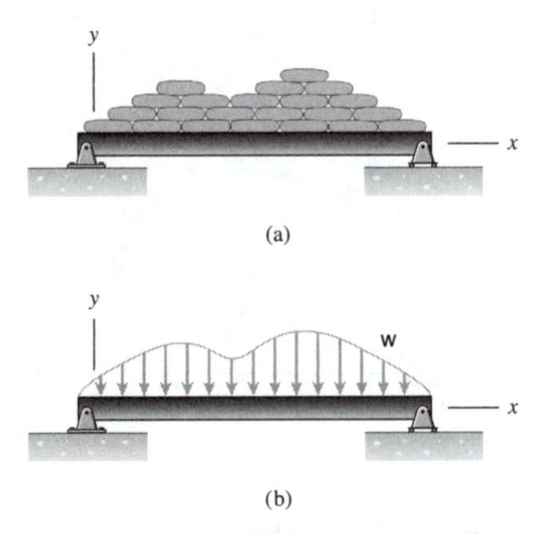

(a) Loading a beam with bags of sand. (b) The distributed load w models the load exerted by the bags.

- The force exerted by a distributed load is

$$F = \int_L w \, dx.$$

- The moment about the origin due to a distributed load is

$$\int_L xw \, dx.$$

- When you are concerned only with the *total force and moment* exerted by a distributed load, you can represent it by a single equivalent force F provided the distributed load acts at the centroid of the "area" between the loading curve (described by the function $w(x)$) and the *x-axis*.

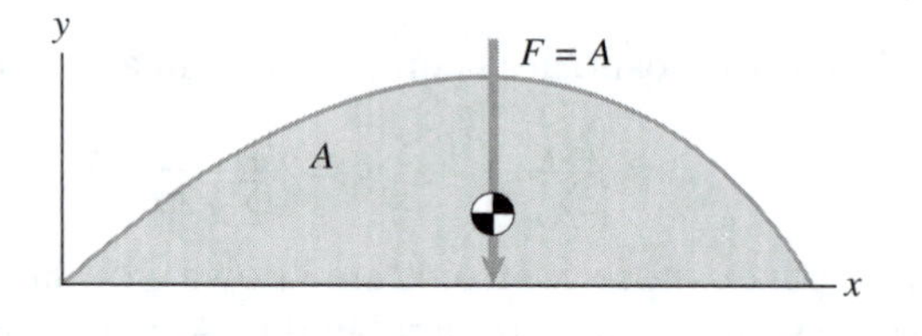

The equivalent force is equal to the "area," and the line of action passes through its centroid.

7.4 Centroids of Volumes and Lines

Here, we define the centroids, or average positions, of volumes and lines, and note how to determine the centroids of composite volumes and lines. Knowing the centroids of volumes and lines allows you to determine the centers of mass of certain types of objects, which tells you where their weights effectively act (see Section 7.7).

- *Definitions*
  - **Volume.** Consider a volume V and let dV be a differential element of V with coordinates x, y, and z. The coordinates of the centroid of V are

$$\bar{x} = \frac{\int_V x dV}{\int_V dV}, \quad \bar{y} = \frac{\int_V y dV}{\int_V dV}, \quad \bar{z} = \frac{\int_V z dV}{\int_V dV}. \tag{7.15}$$

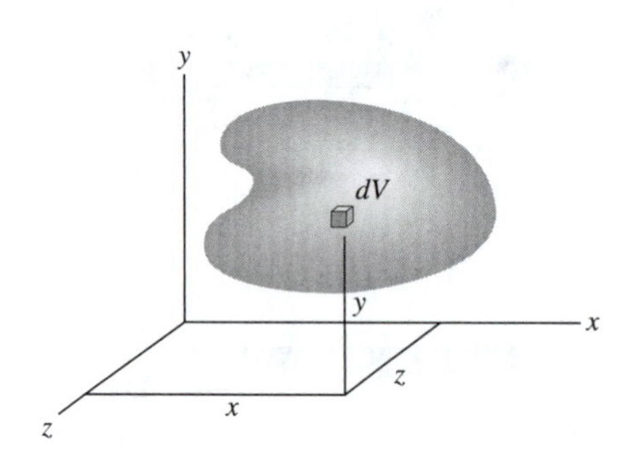

A volume V and differential element dV.

— **Line.** The coordinates of the centroid of a line L are

$$\bar{x} = \frac{\int_L x\,dL}{\int_L dL}, \quad \bar{y} = \frac{\int_L y\,dL}{\int_L dL}, \quad \bar{z} = \frac{\int_L z\,dL}{\int_L dL}. \tag{7.16}$$

where dL is a differential length of the line with coordinates x, y and z.

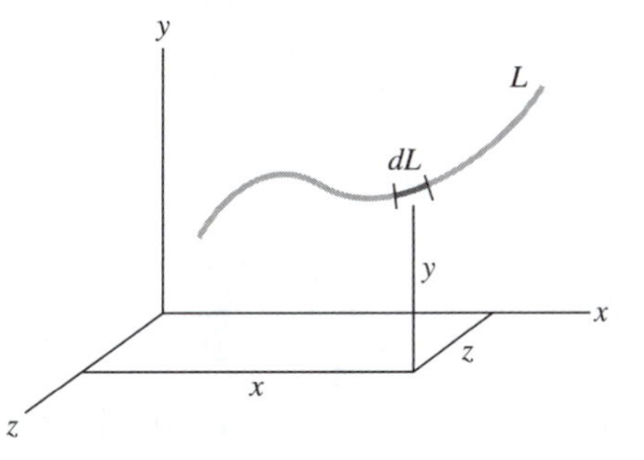

A line L and differential element dL.

7.5 Centroids of Composite Volumes and Lines

The centroids of composite volumes and lines can be derived using the same approach we applied to areas. The coordinates of the centroid of a *composite volume* are

$$\bar{x} = \frac{\sum_i \bar{x}_i V_i}{\sum_i V_i}, \quad \bar{y} = \frac{\sum \bar{y}_i V_i}{\sum_i V_i}, \quad \bar{z} = \frac{\sum \bar{z}_i V_i}{\sum_i V_i}, \tag{7.17}$$

and the coordinates of the centroid of a *composite line* are

$$\bar{x} = \frac{\sum_i \bar{x}_i L_i}{\sum_i L_i}, \quad \bar{y} = \frac{\sum \bar{y}_i L_i}{\sum_i L_i}, \quad \bar{z} = \frac{\sum \bar{z}_i L_i}{\sum_i L_i} \tag{7.18}$$

The centroids of some simple volumes and lines are tabulated in Appendices B and C in the text.

Procedure for Determining the Centroid of a Composite Volume or Line

1. *Choose the parts*—Try to divide the composite into parts whose centroids you know or can easily determine.
2. *Determine the values for the parts*—Determine the centroid and the volume or length of each part. Watch for instances of symmetry that can simplify your task.
3. *Calculate the centroid*—Use Equations (7.17) or (7.18) to determine the centroid of the composite volume or line, respectively.

- Look at Examples 7.11–7.13 in the text which illustrate the procedure.

7.6 The Pappus–Guldinus Theorems

The following two theorems (of Pappus and Guldinus) are used to find the *surface area and volume* of any object of revolution:

- **First Theorem—Surface Area.** Consider a line of length L in the $x-y$ plane with centroid $(\overline{x}, \overline{y})$. The area A of the surface generated by revolving the line about the *x-axis* is

$$A = 2\pi \overline{y} L. \tag{7.19}$$

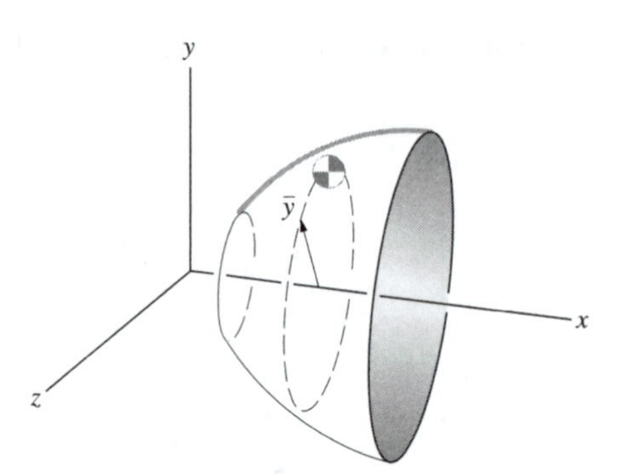

The surface generated by revolving the line L about the x-axis and the path followed by the centroid of the line.

- **Second Theorem—Volume.** Let A be an area in the x-y plane with centroid $(\overline{x}, \overline{y})$. The volume V generated by revolving A about the x-*axis* is

$$V = 2\pi \overline{y} A. \tag{7.21}$$

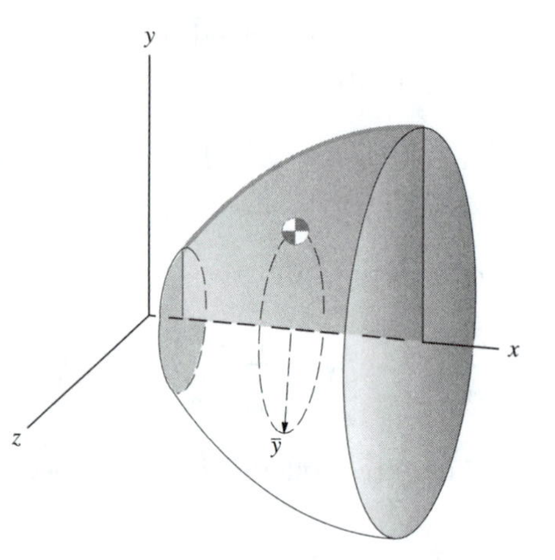

The volume generated by revolving the area A about the x-axis and the path followed by the centroid of the area.

7.7 Centers of Mass of Objects

Background

- The center of mass of an object is the *centroid of its mass* and is defined by

$$\overline{x} = \frac{\int_m x\,dm}{\int_m dm}, \quad \overline{y} = \frac{\int_m y\,dm}{\int_m dm}, \quad \overline{z} = \frac{\int_m z\,dm}{\int_m dm}, \tag{7.23}$$

where x, y, and z are the coordinates of the differential element of mass dm.
- *The weight of an object can be represented by a single equivalent force acting at its center of mass.*

To apply equations (7.23) to specific objects, we change the variable of integration from mass to volume by introducing the mass density.

- The mass density ρ of an object is defined such that the mass of a differential element of its volume is $dm = \rho dV$. The dimensions of ρ are $\frac{\text{(mass)}}{\text{(volume)}}$.
- The total mass of an object is

$$m = \int_m dm = \int_V \rho \, dV. \tag{7.24}$$

- An object whose mass density is *uniform* throughout its volume is said to be *homogeneous.* In this case, the total mass is

$$m = \rho \int_V dV = \rho V. \tag{7.25}$$

- The weight density $\gamma = g\rho$ (N/m^3 or lb/ft^3). The weight of an element of volume dV of an object is $dW = \gamma dV$, and the total weight of a homogeneous object equals γV.

Coordinates of Center of Mass

- The coordinates of the center of mass in terms are:

$$\overline{x} = \frac{\int_V \rho x dV}{\int_V \rho dV}, \quad \overline{y} = \frac{\int_V \rho y dV}{\int_V \rho dV}, \quad \overline{z} = \frac{\int_V \rho z dV}{\int_V \rho dV}. \tag{7.26}$$

It is also worth noting that

- The center of mass of a homogeneous object coincides with the centroid of its volume (just let $\rho = const$ in (7.26)).
- The center of mass of a homogeneous plate of uniform thickness coincides with the centroid of its cross-sectional area.
- The center of mass of a homogeneous slender bar of uniform cross-sectional area coincides approximately with the centroid of the axis of the bar.

7.8 Centers of Mass of Composite Objects

- The coordinates of the center of mass of a composite object composed of parts with masses $m_1, m_2, \ldots$, are

$$\overline{x} = \frac{\sum_i \overline{x}_i m_i}{\sum_i m_i}, \quad \overline{y} = \frac{\sum \overline{y}_i m_i}{\sum_i m_i}, \quad \overline{z} = \frac{\sum \overline{z}_i m_i}{\sum_i m_i} \tag{7.27}$$

where $\overline{x}_i, \overline{y}_i, \overline{z}_i$ are the coordinates of the centers of mass of the parts.
- Since the weights of the parts are related to their masses by $W_i = gm_i$, we can also write Eqs. (7.7) as:

$$\overline{x} = \frac{\sum_i \overline{x}_i W_i}{\sum_i W_i}, \quad \overline{y} = \frac{\sum \overline{y}_i W_i}{\sum_i W_i}, \quad \overline{z} = \frac{\sum \overline{z}_i W_i}{\sum_i W_i}. \tag{7.28}$$

When you know the masses or weights and the centers of mass of the parts of a composite object, you can use these equations to determine its center of mass.

Procedure for Determining the Center of Mass of a Composite Object

1. *Choose the parts*—Try to divide the object into parts whose centers of mass you know or can easily determine.
2. *Determine the values for the parts*—Determine the center of mass and the mass or weight of each part. Watch for instances of symmetry that can simplify your task.
3. *Calculate the center of mass*—Use Equations (7.27) or (7.28) to determine the center of mass of the composite object.

- Look at Examples 7.18–7.19 in the text which illustrate the procedure.

Helpful Tips and Suggestions

- In general, finding the location of the center of mass or centroid involves multiple integration (e.g., over volumes or surfaces). These integrations can often be reduced to single integrations, as in Example 7.8 in the text. Use this or similar procedures (or tables, as in Appendices B and C of the text) when available.

Review Questions

1. True or False?
   (i) The center of mass of an object coincides with the centroid of its volume.
   (ii) The centroid of an object is always located on the object in question.
2. What is the definition of the function w?
3. What is the definition of the *loading curve*?
4. How are the force and moment about the origin exerted by a distributed load determined from the loading curve?
5. If the total weight of an object is to be represented by a single equivalent force, where must this force act?
6. How is the mass density of an object defined?
7. How is the weight density of an object defined?
8. What does it mean when we say a body is homogeneous?
9. If an object is homogeneous, what do you know about the position of its center of mass?
10. What are the theorems of Pappus and Guldinus used for?
11. Show that the centroid for the volume of a body coincides with the *center of mass* only if the material composing the body is *homogeneous*.

8

Moments of Inertia

Main Goals of this Chapter:

- To show how to calculate the moments of inertia of simple areas and objects.
- To use results called *parallel-axis theorems* to calculate moments of inertia of more complex areas and objects.

Areas

8.1 Definitions

Consider an area A in the $x-y$ plane. Four moments of inertia of A are defined

- **Moment of inertia of the area A about the x-axis**:

$$I_x = \int_A y^2 dA.$$

This moment of inertia is sometimes expressed in terms of the *radius of gyration* about the *x-axis*, k_x, defined by

$$I_x = k_x^2 A.$$

- **Moment of inertia of the area A about the y-axis**:

$$I_y = \int_A x^2 dA.$$

This moment of inertia is sometimes expressed in terms of the *radius of gyration* about the *y-axis*, k_y, defined by

$$I_y = k_y^2 A.$$

- **Product of inertia:**

$$I_{xy} = \int_A xy\,dA.$$

Note that if an area is *symmetric* about either the *x-axis* or *y-axis*, its *product of inertia is zero.*

- **Polar moment of inertia:**

$$J_O = \int_A r^2 dA,$$

where $r = \sqrt{x^2 + y^2}$ is the radial distance from the origin of the coordinate system to the dA. The radius of gyration about the origin, k_O, is defined by

$$J_O = k_O^2 A.$$

Note that

$$J_O = I_x + I_y \text{ and } k_O^2 = k_x^2 + k_y^2.$$

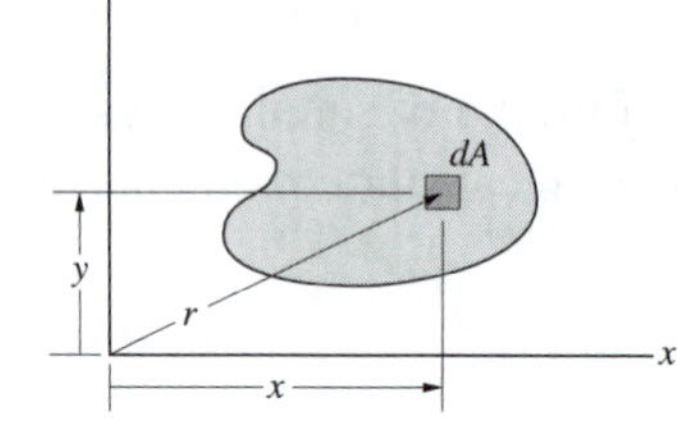

A differential element of A.

- Clearly I_x, I_y and J_O are always positive and have units of length raised to the fourth power.
- The terminology "moment of inertia" is actually a misnomer in this context—it has been adopted because of the similarity with integrals of the same form related to *mass*.

8.2 Parallel-Axis Theorems

In some situations, the moments of inertia of an area are known in terms of a particular coordinate system but we need their values in terms of a different coordinate system. When the coordinate systems are parallel, the desired moments of inertia can be obtained by using the theorems we describe in this section.

- Let $x'y'$ be a coordinate system with its origin at the centroid of an area A and let xy be a parallel coordinate system. The moments of inertia of A in terms of the two systems are related by the *parallel-axis theorems*:

$$I_x = I_{x'} + Ad_y^2,$$

$$I_y = I_{y'} + Ad_x^2,$$

$$I_{xy} = I_{x'y'} + Ad_x d_y,$$

$$J_O = J_O' + Ad^2,$$

where d_x and d_y are the coordinates of the centroid of A in the xy coordinate system.

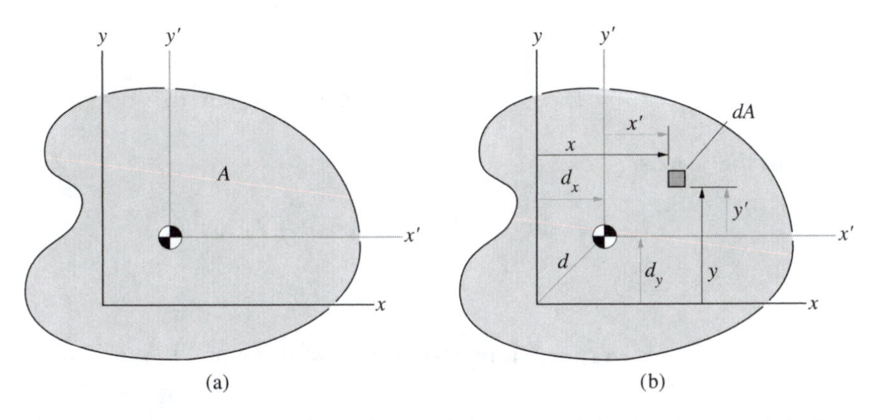

(a) The area A and the coordinate systems $x'y'$ and xy. (b) The differential element dA.

- The parallel-axis theorems can be used to determine the moments of inertia of a composite area:

Procedure for Determining a Moment of Inertia of a Composite Area in Terms of a Given Coordinate System

1. *Choose the parts*—Try to divide the composite area into parts whose moments of inertia you know or can easily determine.
2. *Determine the moments of inertia of the parts*—Determine the moment of inertia of each part in terms of a parallel coordinate system with its origin at the centroid of the part, and then use the parallel-axis theorem to determine the moment of inertia in terms of the given coordinate system.
3. *Sum the results*—Sum the moments of inertia of the parts (or subtract in the case of a hole or a cutout) to obtain the moment of inertia of the composite area.

- Look at Examples 8.3–8.5 in the text which illustrate the procedure.

8.3 Rotated and Principal Axes

Rotated Axes

Consider an area A, a coordinate system xy, and a second coordinate system $x'y'$ that is rotated through an angle θ relative to the xy coordinate system.

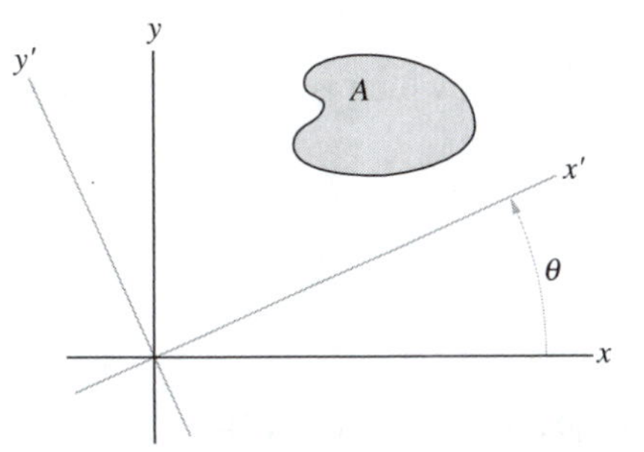

The $x'y'$ coordinate system is rotated through an angle θ relative to the xy coordinate system.

Suppose we know the moments of inertia of A in terms of the xy coordinate system. The moments of inertia in terms of the $x'y'$ coordinate system are given by the following relations:

- **Moment of inertia about the x' axis.**

$$I_{x'} = I_x \cos^2\theta - 2I_{xy}\sin\theta\cos\theta + I_y \sin^2\theta. \tag{8.20}$$

- **Moment of inertia about the y' axis.**

$$I_{y'} = I_x \sin^2\theta + 2I_{xy}\sin\theta\cos\theta + I_y \cos^2\theta. \tag{8.21}$$

- **Product of inertia.**

$$I_{x'y'} = (I_x - I_y)\sin\theta\cos\theta + \left(\cos^2\theta - \sin^2\theta\right) I_{xy}. \tag{8.22}$$

- **Polar moment of inertia.**

$$J'_O = I_{x'} + I_{y'} = I_x + I_y = J_O.$$

Thus, the value of the polar moment of inertia is unchanged by a rotation of the coordinate system.

Principal Axes

- A rotated coordinate system $x'y'$ that is oriented so that $I_{x'}$ and $I_{y'}$ have maximum or minimum values is called a set of *principal axes* of the area A. The corresponding moments of inertia $I_{x'}$ and $I_{y'}$ are called the *principal moments of inertia*.
- The product of inertia $I_{x'y'}$ corresponding to a set of principal axes equals zero.
- Determining principal axes and principal moments of inertia of an area involves three steps:
  1. **Determine I_x, I_y and I_{xy}.**
  2. **Determine the angle θ_P** from the equation

  $$\tan 2\theta_P = \frac{2I_{xy}}{I_y - I_x}. \tag{8.26}$$

  This will determine the orientation of the principal axes within an arbitrary multiple of 90°.
  3. **Calculate $I_{x'}$ and $I_{y'}$.** Once you have chosen the orientation of the principal axes, use Eqs. (8.20)–(8.21) to determine the principal moments of inertia.

8.4 Mohr's Circles

Equations (8.20)–(8.22) have a graphical solution which is convenient to use and easy to remember. This solution is called a *Mohr's circle.*

Masses

Background

- The mass moment of inertia of an object about an axis L_O is

$$I_O = \int_m r^2 dm \tag{8.27}$$

where r is the perpendicular distance from L_O to the differential element of mass dm.

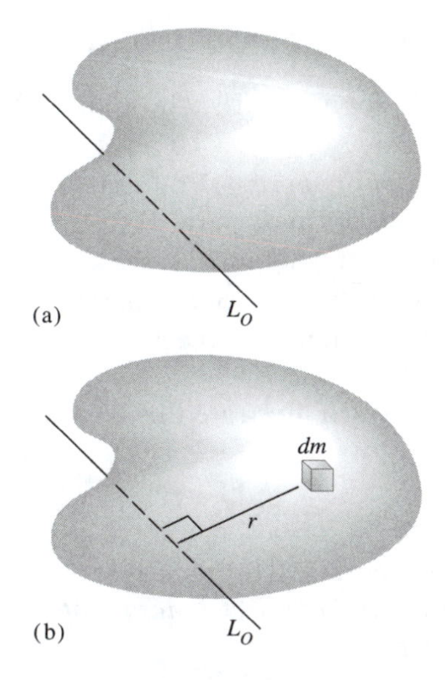

(a) An object and axis L_O. (b) A differential element of mass dm.

8.5 Simple Objects

- The mass moment of inertia of *complicated objects* can be determined by summing the mass moments of inertia of their individual parts (usually *simple objects*). Mass moments of inertia of *simple objects* can be calculated using the formula (8.27). See Appendix C of the text.

8.6 Parallel-Axis Theorem

The *parallel-axis theorem* allows us to determine the mass moment of inertia of an object about any axis when the mass moment of inertia about a parallel axis through the center of mass is known. This theorem can be used to calculate the mass moment of inertia of a composite object about an axis given the mass moments of inertia of each of its parts about parallel axes.

- Let L be an axis through the center of mass of an object, and let L_O be a parallel axis. The moment of inertia I_O about L_O is given in terms of the moment of inertia I about L by the parallel-axis theorem:

$$I_O = I + d^2 m, \tag{8.35}$$

  where m is the mass of the object and d is the perpendicular distance between L and L_O.

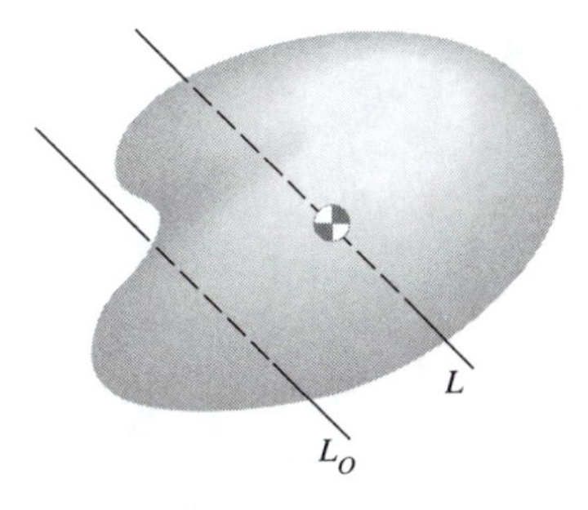

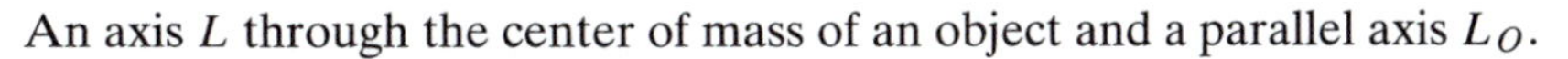

An axis L through the center of mass of an object and a parallel axis L_O.

Procedure for Determining the Mass Moment of Inertia About a Given Axis L_O

1. *Choose the parts*—Try to divide the object into parts whose mass moments of inertia you know or can easily determine.
2. *Determine the mass moments of inertia of the parts*—You must first determine the mass moment of inertia of each part about the axis through its center of mass parallel to L_O. Then you can use the parallel-axis theorem (8.35) to determine its mass moment of inertia about L_O.
3. *Sum the results*—Sum the mass moments of inertia of the parts (or subtract in the case of a hole or cutout) to obtain the mass moment of inertia of the composite object.

- Look at Examples 8.11–8.13 in the text which illustrate the procedure.

Helpful Tips and Suggestions

- Be careful when using the parallel-axis theorem. It is applicable only to the situation when the moment of inertia of the body about an axis *passing through the object's mass center* is known. It *cannot* be applied in the form $I_O = I_B + md^2$ where B is an arbitrary point.

Review Questions

1. True or False? The area moments of inertia I_x, I_y and J_O can be negative.
2. It is often the case that the moment of inertia for an area is known about an axis passing through its centroid. What can then be said about the moment of inertia of the same area about a corresponding parallel axis?
3. How are the radii of gyration for a planar area determined?
4. If a composite area has a "hole", how would you find its moment of inertia ?
5. What are the *principal axes of an area* and the *principal moments of inertia*?
6. What can be said about the product of inertia $I_{x'y'}$ corresponding to a set of principal axes?
7. What are *Mohr's circles* used for?
8. Define the mass moment of inertia and describe how it can be found.
9. What's the parallel-axis theorem for mass moments of inertia?

9

Friction

Main Goals of this Chapter:

- To introduce the concept of dry friction and show how to analyze the equilibrium of objects subjected to this force.
- To present specific applications of frictional force analysis on wedges, threads, bearings, and belts.

9.1 Theory of Dry Friction

Coefficients of Friction

- The theory of *dry friction,* or *Coulomb friction*, predicts the maximum friction forces that can be exerted by dry, contacting surfaces that are *stationary* relative to each other. It also predicts the friction forces exerted by the surfaces when they are in relative motion or *sliding*.
- **Surfaces not in relative motion—THE STATIC COEFFICIENT**
  - ♦ The magnitude of the *maximum* friction force that can be exerted between two plane dry surfaces in contact is

$$f = \mu_s N, \tag{9.1}$$

  where N is the normal component of the contact force between the surfaces and μ_s is a constant called the *coefficient of static friction*.
  - ♦ The value of μ_s is assumed to depend only on the materials of the contacting surfaces and the conditions (smoothness and degree of contamination by other materials) of the surfaces. Typical values of μ_s are given in the following table:

Table 9.1: Typical values of the coefficient of static friction

Materials	**Coefficient of Static Friction μ_s**
Metals on metal	0.15–0.20
Masonry on masonry	0.60–0.70
Wood on wood	0.25–0.50
Metal on masonry	0.30–0.70
Metal on wood	0.20–0.60
Rubber on concrete	0.50–0.90

- Since the friction force in Equation (9.1) is a maximum, Equation (9.1) is applicable when the two surfaces are on the verge of slipping relative to each other. We say that *slip is impending* and the friction forces *resist the impending motion*. In other words,

If slip is impending, the magnitude of the friction force is $f = \mu_s N$ and its direction opposes the impending slip.

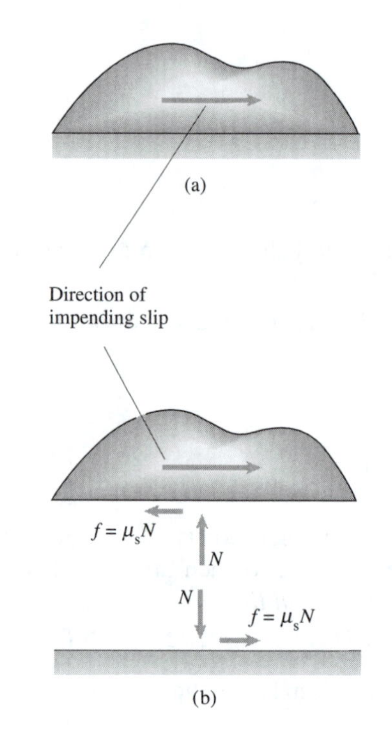

(a) The upper surface is on the verge of slipping to the right. (b) Directions of the friction forces.

- **Surfaces in relative motion—THE KINETIC COEFFICIENT**
  - The magnitude of the friction force between two plane dry contacting surfaces that are in motion (sliding) relative to each other is

$$f = \mu_k N, \tag{9.2}$$

where N is the normal force between the surfaces and μ_k is a constant called the *coefficient of kinetic friction*.

- The value of μ_k is assumed to depend only on the materials of the contacting surfaces and their conditions.
- For a given pair of surfaces, in general, $\mu_k < \mu_s$,

If surfaces are sliding relative to each other, the magnitude of the friction force is $f = \mu_k N$ and its direction opposes the relative motion.

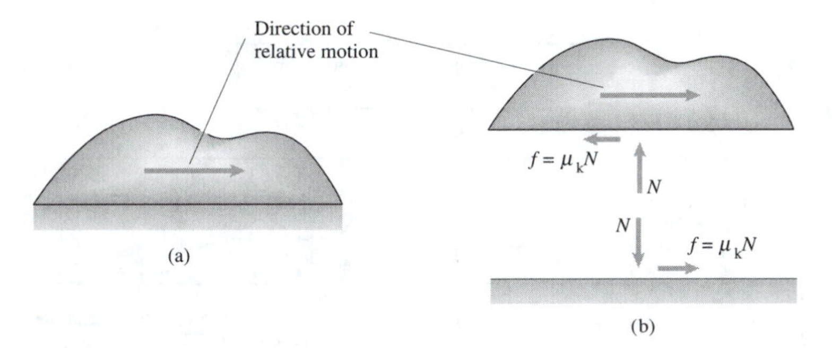

(a) The upper surface is moving to the right relative to the lower surface. (b) Directions of the friction forces.

Angles of Friction

- Instead of resolving the reaction exerted on a surface due to its contact with another surface into the normal force N and friction force f, we can express it in terms of its magnitude R and the angle of friction θ between the force and the normal to the surface.

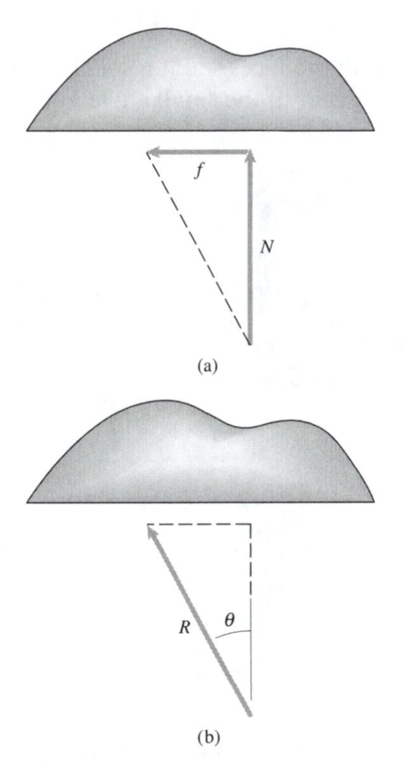

(a) The friction force f and the normal force N. (b) The magnitude R and the angle of friction θ.

- The value of θ when *slip is impending* is called the angle of static friction θ_s, and its value when the surfaces are *sliding relative to each* other is called the angle of kinetic friction θ_k. We have

$$\tan\theta_s = \mu_s,$$

$$\tan\theta_k = \mu_k.$$

- The sequence of decisions in evaluating the friction force and angle of friction is summarized as follows:

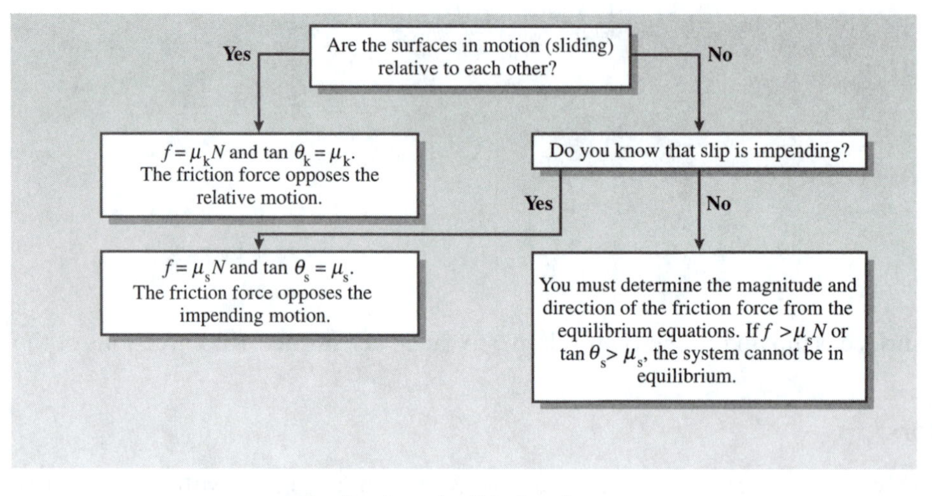

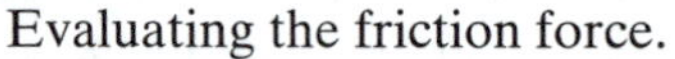
Evaluating the friction force.

9.2 Wedges

- Wedges—A *wedge* is a bifacial tool with the faces set at a small acute angle. When a wedge is pushed forward, the faces exert large lateral forces as a result of the small angle between them.

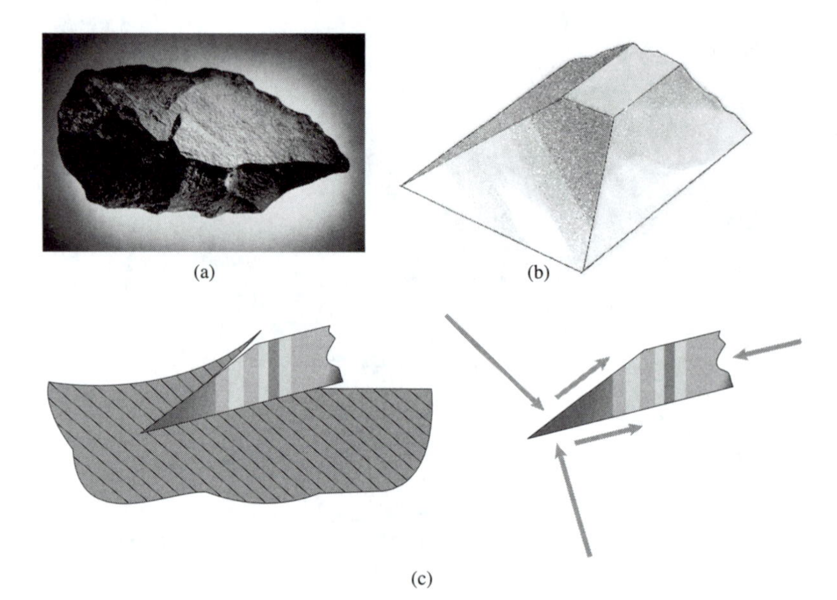

(a) An early wedge tool—a bifacial "hand axe" from Olduvai Gorge, Tanzania. (b) A modern chisel blade. (c) The faces of a wedge can exert large lateral forces.

9.3 Threads

♦ The slope α of the thread is related to its pitch p by

$$\tan\alpha = \frac{p}{2\pi r}$$

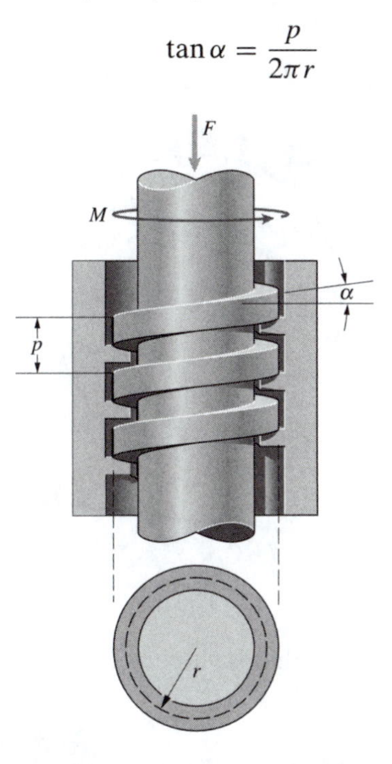

♦ The couple required for impending rotation and axial motion opposite to the direction of F is

$$M = rF\tan(\theta_s + \alpha).$$

♦ The couple required for impending rotation and axial motion of the shaft in the direction of F is

$$M = rF\tan(\theta_s + \alpha).$$

♦ When $\theta_s < \alpha$ the shaft will rotate and move in the direction of the force F with no couple applied.

9.4 Journal Bearings

♦ A *bearing* is a support. This term usually refers to supports designed to allow the supported object to move. For example, in the figure, the circular shaft is supported by two *journal bearings* which allow the shaft to rotate. This shaft can then be used to support a load perpendicular to its axis, such as that subjected by a pulley. The *journal bearings* consist of brackets with holes through which the shaft passes.

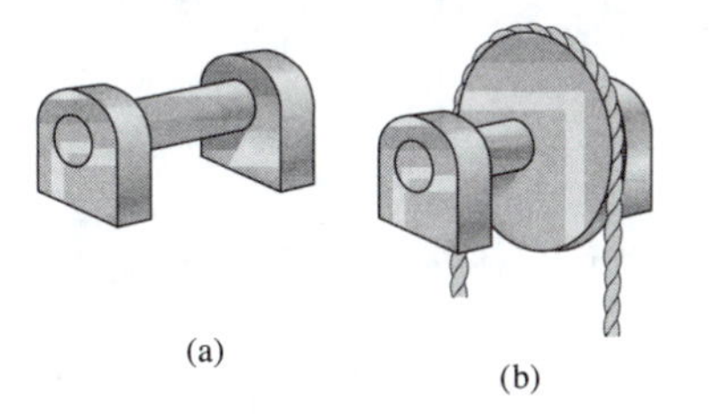

(a) A shaft supported by journal bearings. (b) A pulley supported by the shaft.

♦ The couple required for impending slip of the circular shaft is

$$M = rF\sin\theta_s,$$

where F is the total load on the shaft.

9.5 Thrust Bearings and Clutches

♦ A *thrust bearing* supports a rotating shaft that is subjected to an axial load.

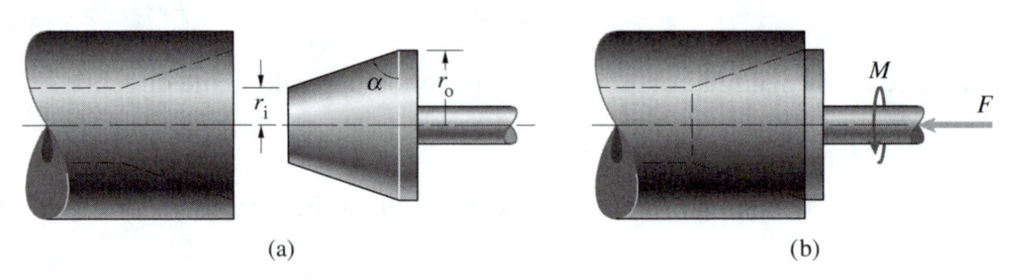

(a) A thrust bearing supports a shaft subjected to an axial load. (b) The differential element dA and the uniform pressure p exerted by the cavity.

♦ A *clutch* is a device used to connect and disconnect two coaxial rotating shafts.

Disengaged position.

♦ The couple required to rotate the shaft at a constant rate is

$$M = \frac{2\mu_k F}{3\cos\alpha}\left(\frac{r_o^3 - r_i^3}{r_o^2 - r_i^2}\right).$$

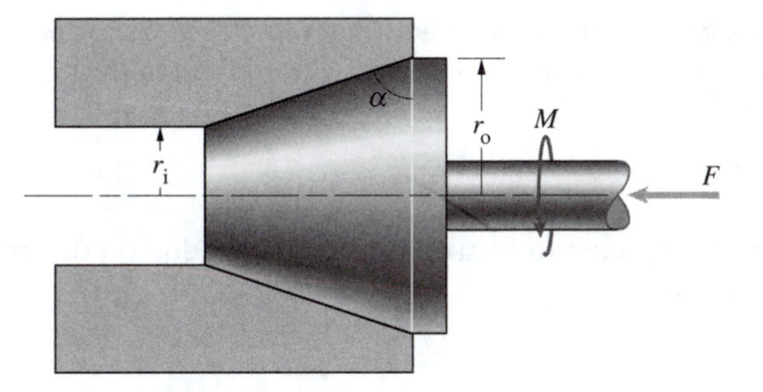

9.6 Belt Friction

- The force T_2 required for impending slip in the direction of T_2 is

$$T_2 = T_1 e^{\mu_s \beta},$$

where β is in radians.

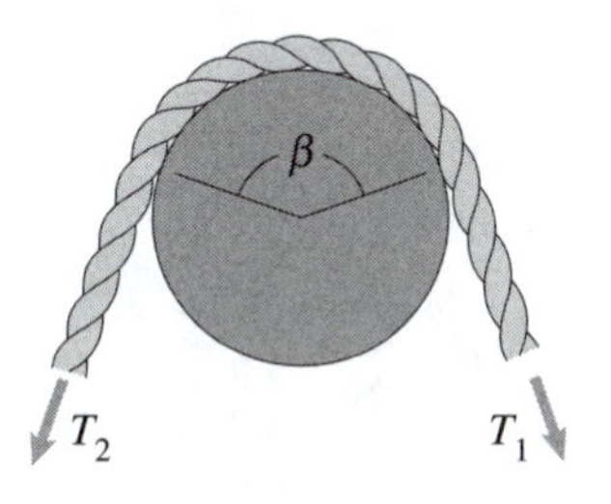

Helpful Tips and Suggestions

- Use *clear* and concise free-body diagrams.
- As in most mechanics problems, *practice* is the key. Make sure you read Examples 9-1 through 9-10 in the text before you attempt any corresponding problems. These examples will serve as templates with which to solve problems. Draw the requested free-body diagrams *yourself.* When doing so, make sure the work is neat and that all the forces and couple moments are properly labelled.

Review Questions

1. How is the coefficient of static friction defined?
2. How is the coefficient of kinetic friction defined?
3. If relative slipping of two dry surfaces in contact is impending, what can you say about the frictional forces they exert on each other?
4. If two dry surfaces in contact are sliding relative to each other, what can you say about the frictional forces they exert on each other?
5. How is the slope α of a thread related to its pitch p?

6. If a threaded shaft is subjected to a large axial load, what is the equation which will give the couple necessary to rotate the shaft at *a constant rate* and cause it to move in the direction of the axial load?
7. What is a *journal bearing*?
8. What is a *thrust bearing?*
9. What is a *clutch*?
10. When the axis of a clutch is subjected to an axial force F, how do you determine the couple M necessary to rotate the shaft at a constant rate?

10

Internal Forces and Moments

Main Goals of this Chapter:

- To use free-body diagrams of parts of individual objects to determine *internal* forces and moments.
- To analyze the forces and moments in a beam subjected to external load and reactions.
- To analyze the forces and study the geometry of cables supporting a load.
- To analyze the force exerted by the pressure of a gas or liquid.

BEAMS

10.1 Axial Force, Shear Force, and Bending Moment

The design of any structural or mechanical member requires an investigation of **both** the external loads and reactions acting on the member **and** the loading acting *within* the member in order to be sure *the material can resist this loading*. These internal loadings can be determined using the *method of sections*.

The idea is to cut an 'imaginary section' through the member so that the internal loadings (of interest) at the section become external on the free-body diagram of the section.

- Since the system of external loads and reactions on the beam is two-dimensional, we can represent the internal forces and moments by an equivalent system consisting of two components of force and a couple. Consider a beam subjected to external load and reactions. The internal forces and moment in a beam are expressed as follows:

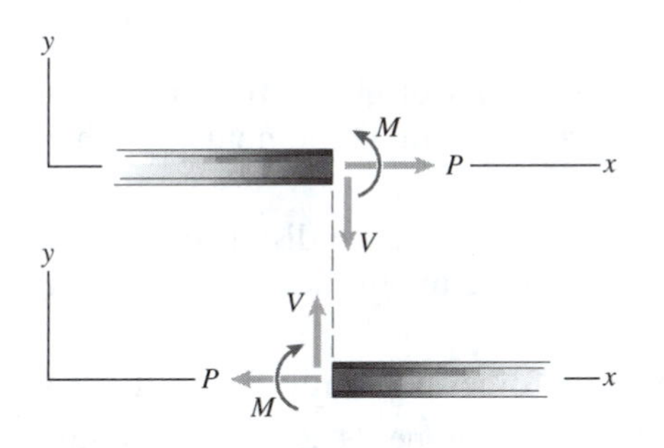

- **Axial force.** This is the component P parallel to the beam's axis.
- **Shear Force.** This is the component V normal to the beam's axis.
- **Bending Moment.** This is the couple M.
- **Directions.** The positive directions of P, V and M are defined in the figure.

PROCEDURE FOR FINDING THE INTERNAL LOADINGS AT A PARTICULAR CROSS SECTION OF A BEAM

1. **Determine the external forces and moments.**
   - Draw the free-body diagram of the beam and determine the reactions at its supports.
   - If the beam is a member of a structure, you must analyze the structure.
2. **Draw the free-body diagram.**
   - Draw the free-body diagram of part of the beam. Cut the beam at the point at which you want to determine the internal forces and moment and draw the free-body diagram of one of the resulting parts. Choose the part with the simplest free-body diagram.
   - If your cut divides a distributed load, don't represent the distributed load by an equivalent force until after you have obtained your free-body diagram.
3. **Apply the equations of equilibrium.**
   - Use the equilibrium equations to determine P, V and M.

10.2 Shear Force and Bending Moment Diagrams

- Beams are designed to support loads *perpendicular* to their axes. The actual design of a *beam* requires a detailed knowledge of the *variation* of the internal shear force V and bending moment M acting at *each point* along the axis of the beam. After this, the theory of mechanics of materials is used with an appropriate engineering design code to determine the beam's required cross-sectional area.
- The *variations* of V and M as functions of the position x along the beam's axis can be obtained using the method of sections. However, it is necessary to section the beam at an arbitrary distance x from one end rather than at a specified point. Depending on the loadings and supports of the beam, it may be necessary to draw several free-body diagrams to determine the distributions of the entire beam. If the results are plotted, the graphical variations of V and M as functions of x are termed the *shear force* and *bending moment diagrams*, respectively. They permit you to see the changes in shear force and bending moment that occur along the beam's length as well as their maximum and minimum values.

  These diagrams can be constructed as follows:

- **Support Reactions.**
  - Determine all the reactive forces and couple moments acting on the beam and resolve all the forces into components acting perpendicular and parallel to the beam's axis.
- **Shear and Moment Functions.**
  - Specify separate coordinates x having an origin at the *beam's left end* and extending to regions of the beam *between* concentrated forces and/or couple moments, or where there is no discontinuity of distributed loading.
  - Section the beam perpendicular to its axis at each distance x and draw the free-body diagram of one of the segments. Be sure $\mathbf{V}$ and $\mathbf{M}$ are shown acting in their *positive sense* in accordance with the established sign convention:
  - The shear V is obtained by summing forces perpendicular to the beam's axis.
  - The moment M is obtained by summing moments about the sectioned end of the segment.

- **Shear and Moment Diagrams.**
  - Plot the shear diagram (V versus x) and the moment diagram (M versus x). If the computed values of the functions describing V and M are *positive*, the values are plotted above the x axis, whereas *negative* values are plotted below the x axis.
  - Generally, it is convenient to plot the shear and bending-moment diagrams directly below the free-body diagram of the beam. See Example 10.3 in the text.
  - For example,

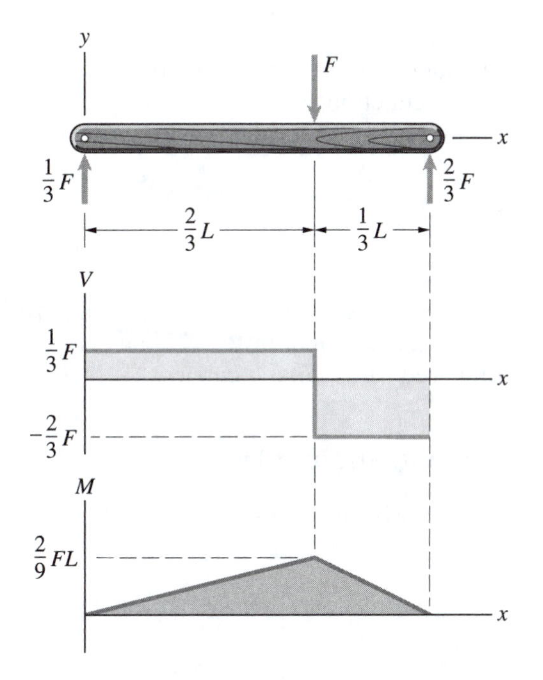

The shear force and bending moment diagrams indicating the maximum and minimum values of V and M.

10.3 Relations Between Distributed Load, Shear Force, and Bending Moment

The shear force V and bending moment M in a beam subjected *only to a distributed load* w are governed by simple differential equations:

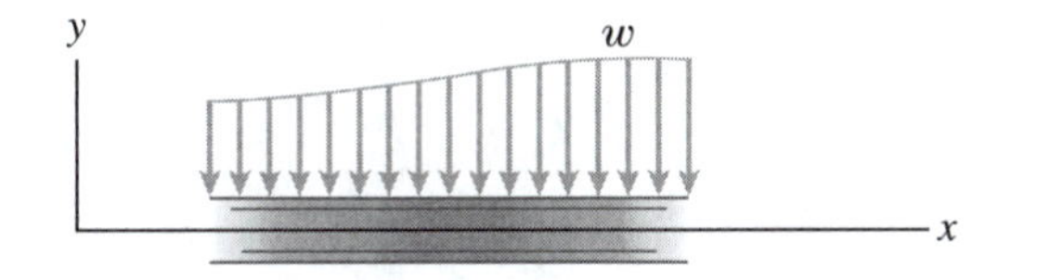

A portion of a beam subjected to a distributed force w.

- The axial force does not depend on x in a portion of a beam subjected only to a lateral distributed load w.

$$\frac{dP}{dx} = 0.$$

- The slope of the shear diagram is equal to the negative of the intensity of the distributed loading, where positive distributed loading is downward i.e.,

$$\frac{dV}{dx} = -w\left(x\right). \tag{10.2}$$

- The slope of the moment diagram is equal to the shear i.e.,

$$\frac{dM}{dx} = V. \tag{10.3}$$

- For segments of a beam that are unloaded or are subjected only to a distributed load w, Eqs. (10.2)–(10.3) can be integrated to determine V and M as functions of x.
- To obtain the complete shear force and bending moment diagrams, forces and couples must also be accounted for.

Cables

Flexible cables and chains are used in engineering structures for support and to transmit loads from one member to another. In the force analysis of such systems, the weight of the cable itself may be neglected (cable is referred to as '*weightless*') because it is often small compared to the load it carries.

10.4 Loads Distributed Uniformly Along Straight Lines

Suppose a suspended cable is subjected to a horizontally distributed load w as follows.

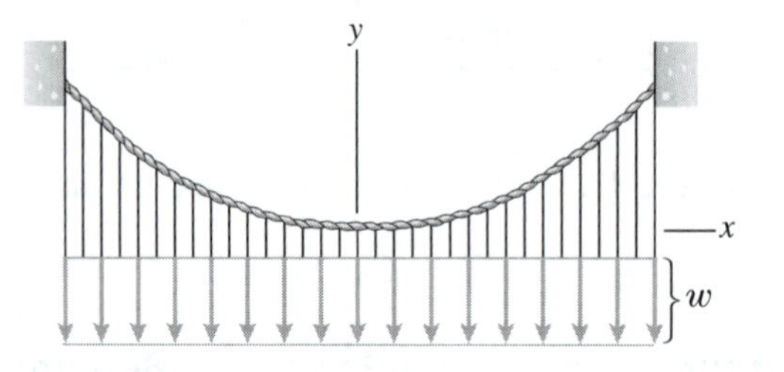

- **Shape of the cable**. The curve described by the cable is the *parabola*

$$y = \frac{1}{2}ax^2,$$

where $a = \dfrac{w}{T_o}$ and T_o is the tension in the cable at $x = 0$.
- **Tension of the cable**. The *tension* in the cable at a position x is

$$T = T_o\sqrt{1 + a^2x^2}.$$

- **Length of the cable**. The length of the cable in the horizontal interval from 0 to x is

$$s = \frac{1}{2}\left\{x\sqrt{1 + a^2x^2} + \frac{1}{a}\ln\left[ax + \sqrt{1 + a^2x^2}\right]\right\}.$$

10.5 Loads Distributed Uniformly Along Cables

A cable's own weight subjects it to a load that is distributed uniformly along its length. If a cable is subjected to equal, parallel forces spaced uniformly along its length, the load on the cable can often be modelled as a load distributed uniformly along its length.

Suppose then that a suspended cable is subjected to a load w distributed along its length. In this case we have the following results.

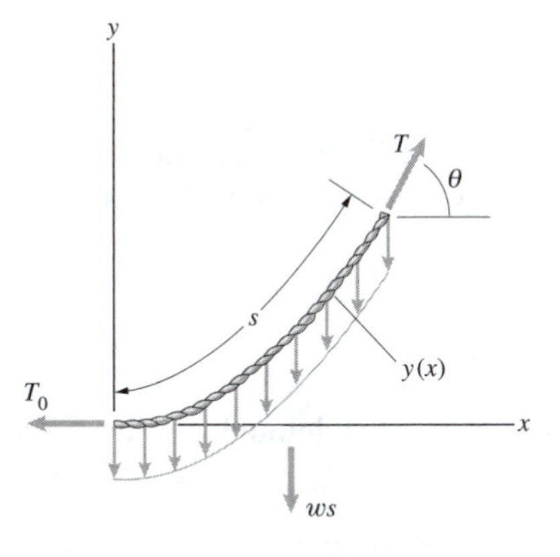

A cable subjected to a load distributed uniformly along its length.

- **Shape of the cable**. The curve described by the cable is the *catenary*

$$y = \frac{1}{2a}\left(e^{ax} + e^{-ax} - 2\right) = \frac{1}{2}\left(\cosh(ax) - 1\right),$$

where $a = \dfrac{w}{T_o}$ and T_o is the tension in the cable at $x = 0$.

- **Tension of the cable**. The *tension* in the cable at a position x is

$$T = T_o\sqrt{1 + \frac{1}{4}\left(e^{ax} - e^{-ax}\right)^2} = T_o \cosh ax.$$

- **Length of the cable**. The length of the cable in the horizontal interval from 0 to x is

$$s = \frac{1}{2a}\left(e^{ax} - e^{-ax}\right) = \frac{\sinh ax}{a}.$$

10.6 Discrete Loads

Consider the case of an arbitrary number N of objects suspended from a cable.

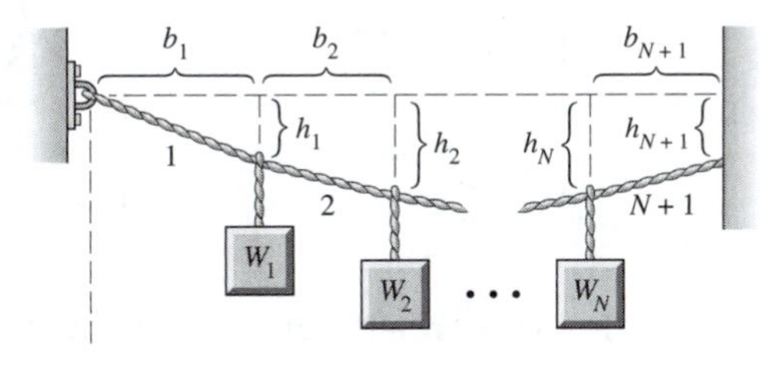

N weights suspended from a cable.

We assume that the weight of the cable can be neglected in comparison to the suspended weights and that the cable is sufficiently flexible that we can approximate its shape by a series of straight segments.

- If N known weights are suspended from a cable (as shown) and the positions of the attachment points of the cable, the horizontal positions of the attachment points of the weights, and the vertical position of the attachment point of one of the weights are known, the configuration of the cable and the tension in each of its segments can be determined.

Liquids and Gases

Many forces of concern in engineering are distributed over areas, for example, wind forces on buildings and aerodynamic forces on cars and airplanes. Here, we analyze the most familiar example, the force exerted by the pressure of a gas or liquid.

10.7 Pressure and the Center of Pressure

- The *pressure* p $\left(\text{units of } \dfrac{force}{area}\right)$ on a surface is defined so that the normal force exerted on an element dA of the surface is $p\,dA$.
- In some applications, it is convenient to use the *gage pressure*

$$p_g = p - p_{\text{atm}},$$

where p_{atm} is the pressure of the atmosphere. Atmospheric pressure varies with location and climatic conditions. Its value at sea level is approximately $1 \times 10^5\,Pa$ in SI units and 14.7 psi or 2120 lb/ft^2 in U.S. Customary units.
- The *total normal force* exerted by pressure on a plane area A is

$$F = \int_A p\,dA.$$

- If the distributed force due to pressure on a surface is represented by an equivalent force F, the point at which the line of action of the force intersects the surface is called the *center of pressure*. In other words, the *center of pressure* is the point on A at which F must be placed to be equivalent to the pressure on A.

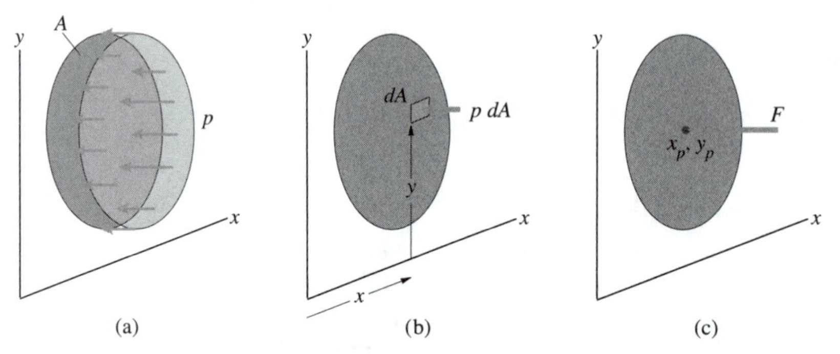

(a) A plane area subjected to pressure. (b) The force on a differential element dA. (c) The total force acting at the center of pressure.

- The *coordinates of the center of pressure* are

$$x_p = \frac{\int_A xpdA}{\int_A pdA}, \quad y_p = \frac{\int_A ypdA}{\int_A pdA}.$$

- It can be shown that the center of pressure coincides with the x and y coordinates of the centroid of the "volume" between the surface defined by the pressure distribution and the area A.

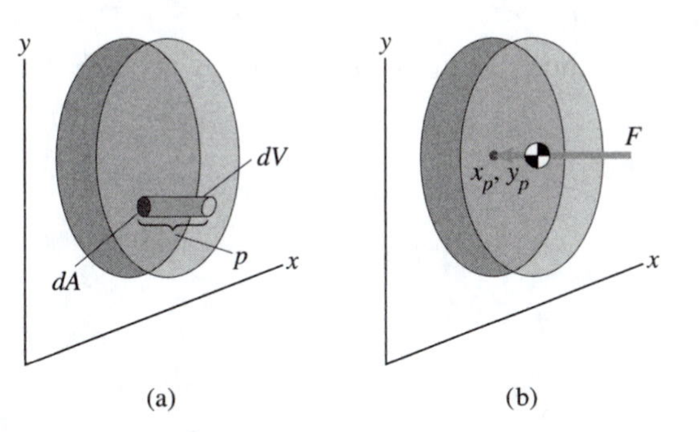

(a) The differential element $dV = p\,dA$. (b) The line of action of F passes through the centeriod of V.

Pressure in a Stationary Liquid

Designers of pressure vessels and piping, ships, dams, and other submerged structures must be concerned with forces and moments exerted by water pressure. The pressure in a liquid at rest increases with depth. In fact,

- The pressure in a stationary liquid is

$$p = p_o + \gamma x,$$

where p_o is the pressure at the surface, γ is the weight density of the liquid, and x is the depth.
- If the surface of the liquid is open to the atmosphere, $p_o = p_{\text{atm}}$, the atmospheric pressure.

Helpful Tips and Suggestions

- Use *clear* and concise free-body diagrams.
- As in most mechanics problems, *practice* is the key. Make sure you read Examples 10.1 through 10.12 in the text before you attempt any corresponding problems. These examples will serve as templates with which to solve problems. Draw the requested free-body diagrams *yourself.* When doing so, make sure the work is neat and that all the forces and couple moments are properly labelled.

Review Questions

1. What are the axial force, shear force and bending moment?
2. How are the positive directions (sense) of the shear force **V** and bending moment **M** defined?

3. For a portion of a beam which is subjected only to a distributed load w, how are the shear force and bending moment distributions determined from equations (10.2) and (10.3)?
4. What does it mean when a cable is assumed to be 'weightless'?
5. If a cable is subjected to a load that is uniformly distributed along a straight line and its weight is negligible, what mathematical curve describes its shape?
6. What is the definition of the pressure p?
7. What is the *gage pressure*?
8. How is the *center of pressure* defined?
9. How can the "volume" defined by the pressure distribution be used to determine the location of the center of pressure?

11

Virtual Work and Potential Energy

Main Goals of this Chapter:

- To define *work* and *potential energy*.
- To introduce the principle of *virtual work* and show how it applies to determining the equilibrium configuration of a series of pin-connected members.

11.1 Virtual Work

Work of a Force

- A force **F** does work only when it undergoes a displacement in the direction of the force.
- Work is a *scalar quantity* defined by the dot product

$$\begin{aligned} dU &= \mathbf{F}.d\mathbf{r} \\ &= F\cos\theta ds, \end{aligned} \tag{11.1}$$

where dU is the increment of work done when the force **F** is displaced $d\mathbf{r}$, θ is the angle between the tails of $d\mathbf{r}$ and **F**, and ds is the magnitude of $d\mathbf{r}$.

- *Positive work* is done when the force and its displacement have the same *sense*. Otherwise *negative work* is done.
- In the *SI* system, the basic unit of work is a *Joule* (J) $(1J = 1N \cdot m)$. In the FPS system work is defined in terms of *ft.lb*.

Work of a Couple

- When an object acted on by a couple M is rotated through an angle $d\alpha$ in the same direction as the couple, the resulting work is

$$dU = M d\alpha. \tag{11.2}$$

If the direction of the couple is opposite to the direction of $d\alpha$, the work is *negative.*

Virtual Work

- A *virtual* movement (displacement or rotation) is an *imaginary* movement which is *assumed* and *does not actually exist*. A *virtual displacement* is a differential that is given in the positive direction of the position coordinate and is denoted by the symbol δs . Similarly, a *virtual rotation* is denoted by $\delta\alpha$.
- The *virtual work* done by a force undergoing a virtual displacement δs is

$$\delta U = F \cos\theta \delta s.$$

- The *virtual work* done by a couple undergoing a virtual rotation $\delta\alpha$ in the plane of the couple forces is

$$\delta U = M\delta\alpha.$$

Principle of Virtual Work

- *If an object is in equilibrium, the virtual work done by the external forces and couples acting on it is zero for any virtual translation or rotation:*

$$\delta U = 0. \tag{11.9}$$

This principle can be used to derive the equilibrium equations for an object. However, there is no advantage to this approach compared to simply drawing the free-body diagram of the object and writing the equations of equilibrium in the usual way. *The advantages of the principle of virtual work become evident when we consider structures.*

Application to Structures

The principle of virtual work applies to each member of a structure. By subjecting certain types of structures in equilibrium to virtual motions and calculating the total virtual work, we can determine unknown reactions at their supports as well as internal forces in their members. The procedure involves finding virtual motions that result in virtual work being done by both known loads and by unknown forces and couples:

Using Virtual Work to Determine Reactions on Members of Structures

1. **Choose a virtual motion**. Identify a virtual motion that results in virtual work being done by known loads and by an unknown force or couple you want to determine.
2. **Determine the virtual work**. Calculate the total virtual work resulting from the virtual motion to obtain an equation for the unknown force or couple.
3. The above procedure is illustrated in Examples 11-1 through 11-2 in the text. *Note that had these examples been solved using the equations of equilibrium, it would have been necessary to dismember the links and apply three scalar equations to each link. The principle of virtual work, by means of calculus, has eliminated this task so that the answer is obtained directly.*

11.2 Potential Energy

- If a function of position V exists such that for any displacement $d\mathbf{r}$, the work done by a force $\mathbf{F}$ is

$$dU = \mathbf{F}.d\mathbf{r} = -dV, \tag{11.10}$$

V is called the potential energy associated with the force, and $\mathbf{F}$ is said to be *conservative*.

Examples of Conservative Forces: Weight, Elastic Springs

- **Gravitational Potential Energy.** The potential energy associated with the weight W of an object is

$$V = Wy \tag{11.12}$$

where y is the height of the center of mass above some reference level, or *datum.*

- **Elastic Potential Energy.** The potential energy associated with the force exerted by a linear spring is

$$V = \frac{1}{2}kS^2, \tag{11.14}$$

where k is the spring constant and S is the stretch of the spring.

- **Potential Function.** In the general case, if a body is subjected to both gravitational and elastic forces, the potential energy (function) V of the body can be expressed as the algebraic sum

$$V = V_g + V_e$$

where measurement of V depends on the location of the body with respect to a selected datum in accordance with Eqs. (11.7) and (11.8).

Examples of Nonconservative Forces

- *Friction:* the work done by the frictional force *depends on the path*. The longer the path, the greater the work. The work done is dissipated from the body in the form of heat.

Principle of Virtual Work for Conservative Forces

- Since the work done by a conservative force is expressed in terms of its potential energy through Eq. (11.6), we can give an alternative statement of the principle of virtual work when an object is subjected to conservative forces:

Suppose that an object is in equilibrium. If the forces that do work as
as the result of a virtual translation or rotation are conservative,
the change in the total potential energy is zero.

In other words,

$$\delta V = 0. \tag{11.15}$$

- **Note** that it is not necessary for *all* of the forces acting on the object be conservative for this result to hold: it is necessary only that the forces that *do work* be conservative.
- This principle also applies to a system of interconnected objects if the external forces that do work are conservative and the internal forces at the connections between objects either do no work or are conservative. Such a system is called a *conservative system*.
- If the position of an object or a system can be specified by a single coordinate q, it is said to have *one degree of freedom*.
- When a conservative, one-degree-of-freedom object or system is in equilibrium,

$$\frac{dV}{dq} = 0.$$

We can use this equation to determine the values of q at which the system is in equilibrium.

Stability of Equilibrium

Once the equilibrium configuration for a body or a system of connected bodies is defined, it is important to investigate the "type" of equilibrium or the stability of the configuration.

- Types of Equilibrium.
  1. *Stable Equilibrium.* A small displacement of the system causes the system to return to its original position. Potential energy of the system is at a minimum in this case.
  2. *Unstable Equilibrium.* A small displacement of the system causes the system to move farther away from its original position. Original potential energy of the system is a maximum in this case.
- System Having One Degree of Freedom (q). If a conservative, one-degree-of-freedom system is in equilibrium, we require that the potential energy (function) V of the body satisfies the following conditions in each case:
  1. *Stable Equilibrium.*

$$\frac{dV}{dq} = 0, \quad \frac{d^2V}{dq^2} > 0. \tag{A}$$

  2. *Unstable Equilibrium.*

$$\frac{dV}{dq} = 0, \quad \frac{d^2V}{dq^2} < 0. \tag{B}$$

Using Potential Energy to Analyze the Equilibrium of One-Degree-of-Freedom Systems

1. **Determine the potential energy**—Express the total potential energy in terms of a single coordinate that specifies the position of the system.
2. **Find the equilibrium positions**—By calculating the first derivative of the potential energy, determine the equilibrium position or positions.
3. **Examine the stability**—Use the sign of the second derivative of the potential energy to determine whether the equilibrium positions are stable.

Helpful Tips and Suggestions

- Remember that work and energy are *scalars*—not vectors—even though they arise from the product of two vectors.
- As in most mechanics problems you must develop a sense of *which* method is most suitable in tackling a particular problem. For example, there is no advantage to the virtual work approach for a single object compared to simply drawing the free-body diagram of the object and writing the equations of equilibrium in the usual way. *The advantages of the principle of virtual work become evident when we consider structures.* Developing the skills necessary to know which method is 'best' comes from *experience and practice.*

Review Questions

1. What is the work done by a force **F** when its point of application is displaced $d\mathbf{r}$?
2. What is the work done by a couple **M** when the object on which it acts rotates through an angle $d\alpha$ in the same direction as the couple?

3. What does the principle of virtual work say when an object in equilibrium is subjected to a virtual translation or rotation?
4. What is the potential energy of a body and how is it related to the concept of "conservative force"?
5. Give an example of a conservative force.
6. If an object in equilibrium is subjected only to conservative forces, what do you know about the total potential energy when the object undergoes a virtual translation or rotation?
7. What does it mean when an equilibrium position of a body is stable or unstable?
8. How do you know when an equilibrium position of a conservative system having one degree of freedom is stable or unstable?

Answers to Review Questions

- Chapter 1:

1. F **2.** T **3.** T **4.** F **5.** T **6.** F **7.** F **8.** T **9.** F **10.** T

- Chapter 2:

1. See (2.7) **2.** See (2.8) **3.** See (2.10)
4. If the fingers of the right-hand are pointed in the positive *x-direction* and then closed toward the positive *y-direction*, the thumb points in the *z-direction*.
5. See (2.15) and (2.16).
6. See (2.18). Scalar.
7. Vectors are perpendicular
8. See (2.23)
9. See (2.20) **10.** See (2.26) and (2.27)

- Chapter 3:

1. A system of forces is *coplanar* or *two-dimensional* if the lines of action of the forces lie in the plane. Otherwise it is *three-dimensional.*
2. If the friction force is negligible in comparison to the normal force, the surfaces are said to be *smooth*.
3. An object is in equilibrium only if each point of the object has the same constant velocity (steady translation).
4. Must be zero.
5. **r** is the position vector from P to *any* point on the line of action of **F**.
6. Lines of action of the forces lie in the plane.
7. Lines of action of the forces lie in three-dimensional space.
8. One more equation—see (3.4) and (3.5).
9. See (3.2).
10. True.

- Chapter 4:

1. See Section 4.1. A vector. **2.** See (4.1) **3.** Right-hand rule.
4. No moment.

5. **r** represents a position vector drawn *from P to any point* lying on the line of action of **F**.
6. See (4.3).
7. Right-hand rule—curling the fingers of the right hand from vector **r** (cross) to vector **F**, the thumb then points in the direction of $\mathbf{M}_P$.
8. See Section 4.4
9. False—same about any point.
10. See Section 4.5.
11. See Section 4.5.
12. A force **F** and a couple $\mathbf{M}_P$ that is parallel to **F**—see Section 4.5.

- Chapter 5:

1. No more than three.
2. When an object has more supports than the minimum number necessary to maintain it in equilibrium.
3. When there are more unknown loadings on the body than equations of equilibrium available for their solution.
4. Yes, from Table 5.2, a built-in or fixed support has already 6 unknowns (three force components and three couple components). Hence, any additional supports will lead to a situation in which there are more unknowns than independent equilibrium equations available (6).
5. From Table 5.2, five: three force components and two couple components.

- Chapter 6:

1. A *truss* is a structure composed entirely of two-force members.
2. See Section 6.2.
3. Two.
4. See Section 6.3.
5. See Section 6.3.
6. Space trusses are analyzed by the same methods described for two-dimensional trusses. The only difference is the need to cope with the more complicated geometry. See Section 6.4.
7. See Section 6.5. Structures of interconnected members that *do not* satisfy the definition of a truss are called *frames* if they are designed to remain stationary and support loads and *machines* if they are designed to move and apply loads.
8. When trusses are analyzed by cutting members to obtain free-body diagrams of joints or sections, the internal forces acting at the "cuts" are simple axial forces. This is not generally true for frames or machines, and a different method of analysis is necessary.
9. Instead of cutting members, you isolate entire members, or in some cases, combinations of members, from the structure.
10. See Section 6.5.

- Chapter 7:

1. (i) F (material comprising the body must be *homogeneous*). (ii) F e.g. centroid of a circular ring.
2. A force distributed along a line is described by a function w $\left(\text{units of } \frac{\textit{force}}{\textit{length}}\right)$, defined such that the force on a differential element dx of the line is $w\,dx$.
3. The graph of w is called the *loading curve*.
4. See Section 7.4.

5. At the centroid of the "area" between the loading curve (described by the function $w(x)$) and the $x-axis$.
6. The mass density ρ of an object is defined such that the mass of a differential element of its volume is $dm = \rho dV$. The dimensions of ρ are $\frac{(\text{mass})}{(\text{volume})}$.
7. The weight density $\gamma = g\rho$ (N/m^3 or lb/ft^3)
8. An object whose mass density is *uniform* throughout its volume is said to be *homogeneous.*
9. The center of mass of a homogeneous object coincides with the centroid of its volume.
10. To find the *surface area and volume* of any object of revolution.
11. From (7.26), the coordinates of the center of mass of a body are given by

$$\bar{x} = \frac{\int_V \rho x dV}{\int_V \rho dV}, \quad \bar{y} = \frac{\int_V \rho y dV}{\int_V \rho dV}, \quad \bar{z} = \frac{\int_V \rho z dV}{\int_V \rho dV}.$$

If the body is homogeneous ρ is constant. Thus we obtain:

$$\bar{x} = \frac{\int_V x dV}{\int_V dV}, \quad \bar{y} = \frac{\int_V y dV}{\int_V dV}, \quad \bar{z} = \frac{\int_V z dV}{\int_V dV}$$

which is exactly the centroid of the volume (see Section 7.4).

- Chapter 8:

1. See Section 8.1: False.
2. See Section 8.2: *It's equal to the moment of inertia about the axis* ***passing through the area's centroid*** *plus the product of the area and the square of the perpendicular distance between the axes.*
3. See Section 8.1.
4. If a composite part has a "hole", its moment of inertia is found by "subtracting" the moment of inertia for the hole from the moment of inertia of the entire part including the hole.
5. See Section 8.3. The *principal axes of the area* identify the orientation of the axes x' and y' about which the moments of inertia $I_{x'}$ and $I_{y'}$ are *maximum or minimum.* (the *principal moments of inertia).*
6. Zero.
7. See Section 8.4: Mohr's circle provides a convenient (graphical) means for transforming I_x, I_y and I_{xy} into the principal moments of inertia.
8. See Equation (8.27) and Section 8.6.
9. See Equation (8.35).

- Chapter 9:

1. See Section 9.1: $\mu_s : f = \mu_s N$.
2. See Section 9.1: $\mu k : f = \mu_k N$.
3. *If slip is impending, the magnitude of the friction force is $f = \mu_s N$ and its direction opposes the impending slip.*
4. *If surfaces are sliding relative to each other the magnitude of the friction force is $f = \mu_k N$ and its direction opposes the relative motion.*
5. $\tan\alpha = \frac{p}{2\pi r}$. See Section 9.3.
6. $M = rF\tan(\theta_s + \alpha)$. See Section 9.3.
7. See Section 9.4.
8. See Section 9.5.

9. A *clutch* is a device used to connect and disconnect two coaxial rotating shafts.

10. $M = \dfrac{2\mu_k F}{3\cos\alpha}\left(\dfrac{r_o^3 - r_i^3}{r_o^2 - r_i^2}\right)$. See Section 9.5.

- Chapter 10:

1. See Section 10.1: Axial force **P** acts parallel to the beam's axis. Shear force **V** acts normal to the beam's axis. Bending moment **M** is a couple moment which causes the beam to bend.

2. See Section 10.1. Follow the sign convention shown in the figure.

3. By integration to obtain V and M as functions of x.

4. The weight of the cable itself may be neglected. No cable is truly 'weightless'. This is a simplification to aid the modeling.

5. See Section 10.4—a parabola.

6. The *pressure* p $\left(\text{units of } \dfrac{force}{area}\right)$ on a surface is defined so that the normal force exerted on an element dA of the surface is $p\,dA$. See Section 10.7.

7. The gage pressure is defined as $p_g = p - p_{atm}$, where p_{atm} is the pressure of the atmosphere.

8. If the distributed force due to pressure on a surface is represented by an equivalent force F, the point at which the line of action of the force intersects the surface is called the *center of pressure*. See Section 10.7.

9. The center of pressure coincides with the x and y coordinates of the centroid of the "volume" between the surface defined by the pressure distribution and the area A. See Section 10.7.

- Chapter 11:

1. See (11.1). **2.** See (11.2).

3. That the virtual work (δU) done by the force system must be zero. See (11.9).

4. See Section 11.2.

5. Weight. See Section 11.2.

6. The change in the total potential energy is zero. See Eq. (11.15).

7. *Stable Equilibrium:* A small displacement of the system causes the system to return to its original position. Potential energy of the system is at a minimum in this case; *Unstable Equilibrium.* A small displacement of the system causes the system to move farther away from its original position. Original potential energy of the system is a maximum in this case.

8. Test using Equation (A) for stability and Equation (B) for instability.